DEVELOPING SUBJECT KNOWLEDGE IN DESIGN AND TECHNOLOGY:

FOOD TECHNOLOGY

DEVELOPING SUBJECT KNOWLEDGE IN DESIGN AND TECHNOLOGY:

FOOD TECHNOLOGY

Edited by

Gwyneth Owen-Jackson at The Open University

in association with

Trentham Books

Stoke on Trent, UK and Sterling, USA

Trentham Books Limited

Westview House
734 London Road
Oakhill
Stoke on Trent
Staffordshire
England ST4 5NP

22883 Quicksilver Drive
Sterling
VA 20166-2012
USA

First published 2001

British Library Cataloguing-in-Publication Data
A catalogue record for this book is available from the British Library

ISBN: 1 85856 245 7

Designed and typeset by Trentham Print Design Ltd., Chester and printed in Great Britain by Cromwell Press Ltd., Wiltshire.

Acknowledgements

This text draws heavily on two current Open University courses.

SK220 – Human Biology and Health, book 3 'Maintaining the Whole'

ST240 – Our Chemical Environment, book 3 'Nutrition and Health'

and we want to thank each of the course teams for generously allowing their text to be used.

We also want to thank Ali Farrell for her contribution on food product design and development and food manufacturing.

Contents

Introduction

The aim of this text is to develop subject knowledge in Food Technology for trainee teachers of Design and Technology. You will be specialising in one of the areas of Design and Technology but the current requirement for Newly Qualified Teachers of Design and Technology is that you can teach your specialist area to post 16 level and have a second area that you can teach to key Stage 3 level. This book is designed to help you develop your knowledge of Food Technology sufficiently to be able to teach it to Key Stage 3 level.

Suggestions are given below as to how you may work through this book, but throughout you are seen as an active learner, engaging with the text and developing your knowledge and understanding.

Within the book elements in the National Curriculum for Design and Technology for England are addressed in particular:

Key Stage 3	
aspects of 1. 'Developing, planning and communicating ideas'	Food product development
aspects of 3. 'Evaluating processes and products'	Food product development
4a. consider physical and chemical properties and working characteristics of a range of common and modern materials	The chemical structure of nutrients Food and nutrition
4b. that materials can be classified according to their properties and working characteristics	The chemical structure of nutrients Food and nutrition
4c. that materials can be combined, processed and finished to create more useful properties and particular aesthetic effects.	Food and nutrition
4d. how multiple copies can be made of the same product	Food manufacturing
Key Stage 4	
1c. design for manufacturing in quantity	Food manufacturing
1d. produce and use detailed working schedules, setting realistic deadlines and identifying critical points	Food manufacturing
3b. devise and apply tests to check the quality of their work at critical points during development	Food product development
3c. ensure that their products are of a suitable quality for intended users	Food product development Food manufacturing

In addition, attention has been paid to meeting the knowledge specified as required to teach Food Technology in *Minimum Competences for students to teach Design and Technology in secondary schools* (Design and Technology Association, 1995).

Who is this book for?

This book is intended to be used as part of initial teacher training for Design and Technology, and may be used in a number of ways, for example:

- Pre-course, or at the very start of your course, you may choose to work through it all in one go
- At stages during your course (to suit your own time and needs) to satisfy subject audit requirements
- During school placement, when particular topics in Food Technology are being taught.

How to use this book

The book has been divided into sections which, worked through sequentially, would provide a comprehensive grounding in all aspects. It begins by looking at consumers food choices, food product design and development, and food manufacturing. There are then chapters on food science which take you through from aspects you may be familiar with, but not have thought through in detail, to a scientific understanding of food and how the body utilises it. Finally it considers the relationship between diet and health. It may be that you already have knowledge in some of these areas so need not study them. However, if you need to develop certain aspects of your knowledge or if you require particular knowledge at a particular time for teaching purposes, then each section can be studied independently.

All the sections incorporate activities and questions to help you to develop a thorough understanding of the topics covered. The book has been designed for self study, but if it is possible you may wish to discuss these activities and questions with your professional tutor, school mentor or a Food Technology teacher, as this will help you to put the knowledge into an appropriate and relevant context. Answers to the questions are given at the back of the book.

This book will give you sufficient knowledge to teach 'Materials and Components' through Food Technology to Key Stage 3 level and will help you to develop confidence in delivering this area of the curriculum. There will be the time to further develop your subject knowledge whilst you are teaching. If, in addition to this subject knowledge, you need to develop your skills of design in relation to food products this is covered in 'Developing Subject Knowledge in Design and Technology – Developing, Planning and Communicating Ideas', a book in this series.

Food Technology, like all areas of Design and Technology, is continuously changing and developing as new scientific discoveries are made and as new products are developed. This means that you will need to constantly keep up-dating your knowledge to ensure that you always have the latest information.

Keeping subject knowledge up-to-date is part of being an effective teacher and means you will always be at the forefront of what is happening in your area. This is an aspect of Design and Technology that makes it challenging, interesting and exciting.

You are recommend to purchase a copy of 'The Manual of Nutrition' (MAFF 1995) to read alongside this book.

Influences on food choice

Introduction

Eating is instinctive, yet we know we have to be selective about what we consume – only certain substances such as vegetable and animal tissue will sustain us. What is so special about these materials? Then again, how specific does our diet have to be? Unlike some animals we are omnivores, but if you were dropped on a desert island would you know what to eat to survive? In fact, the great success of homo sapiens in colonizing this planet is partly due to an ability to adapt eating habits to the food which the immediate environment provides, be it a desert, arctic waste, tropical rain forest or fast-food-infested city.

Activity 1

Think for a moment about your own diet. What foods do you eat regularly, occasionally, never – why is this? List the factors that you think influence your own choice of food.

The environment in which we evolved, over 40,000 years ago, has shaped our nutritional needs. Our ancestors had to find out what they could and could not eat by trial and error, and many sacrificed themselves (accidentally!) in finding out what was safe, whereas others used skills of observation and recording (if only in memory) to ensure that it did not happen to them.

From the original hunter-gatherer society, things began to change about 10,000 years ago when cereal farming began, cattle were domesticated and people became settled. Diet also changed; as well as cereal and meat we began to consume eggs, dairy products, alcoholic beverages and salt.

Today we have a greater choice of foods than ever before; improved transport and storage means that foodstuffs are imported from all over the world. Seasonal variations no longer apply, foods are available all year round and manufacturers are constantly creating new foods to tempt us to buy. Rather than adapting our diet to suit our environment we seem to have control over our environment to ensure that we can choose the diet we want.

So, from the vast selection available, how do people choose what to eat?

Obviously, the availability of food is a major factor. This is partly determined by geographical location. Foods available in India, America and Africa will vary, although this is becoming increasingly less so. Even within one country availability can be affected by location, for example in Britain the availability of food will be partly determined by the whether you live in a city near to a number of large supermarkets or ethnic communities, or whether you live in a small rural village with just one local shop.

Activity 2

Consider where you live and the availability of food. Is your choice restricted in any way? Has the availability of food changed in recent years? If it has, in what way and what do you think are the reasons for this?

Is your access to food different from that of other people that you know? If so, what are the reasons for this?

Location may also affect one's economic ability to purchase, the social and religious influences which abound and the technology available to prepare and produce food. These are discussed below.

Nutritional aspects of food choice

The range of foods we eat is known as our **diet**. The components of food which are eaten, absorbed by the body and produce energy, promote growth and repair of the body, or control these processes are called **nutrients**. Our health is dependent upon the nature and quantities of the foods we eat, and the consumption of what is commonly called a balanced diet. Such a diet contains the full complement of nutrients in amounts appropriate for our individual needs.

Activity 3

Write down one day's intake of food that you would consume, perhaps what you ate yesterday. Do you consider that this is a balanced diet? Does it contain a range of foods from different sources? Does it contain different amounts of a range of foods?

Most of us feel that we have some idea of what constitutes a balanced or a healthy diet. However, it is interesting to note that in the UK our ideas about what constitutes a healthy diet have changed rapidly in the 20th century. At the beginning of the century, the state of the urban poor was highlighted by the poor state of health of the men called up to fight in the Boer War. This sparked national concern, so in 1906 the national school meals service was started. The school meals service increased public awareness of the need for children to eat foods that provided them with the energy and protein they needed for growth. School milk was introduced in the 1930s and doses of cod liver oil were given to children to prevent rickets.

The Second World War meant that there was strict food rationing and, to avoid widespread malnutrition, emphasis was placed on the need for home production of vegetables and consumption of sufficient (but not excess) quantities of milk, butter, meat and bread. The diet was filling, it avoided deficiencies and was available to everyone. There was intense nutritional education of the public to eat a balanced diet that incorporated 'protein foods', 'energy foods (carbohydrates and fats) and 'protective foods' to avoid nutritional deficiencies. Many people now consider the wartime diet to be a healthy one which avoided excess consumption of any one dietary constituent.

By the 1950s people were released from food rationing and began to demand luxury items which they had not been able to obtain in the previous years, for example large amounts of meat, dairy produce, sugar and other sweet foods. Cheap food began to be produced in larger quantities, due to improvements in agricultural technology and intensive animal rearing. The idea that Britain needed food in quantity to avoid deficiency persisted from the pre-war years. It was thought that eventually everyone would be able to choose an enjoyable, balanced and cheap diet. The problems of over-production are now exercising minds world-wide.

As the UK became more affluent, the nation's diet changed to one high in animal fats, salt and sugar and low in starchy and fibrous foods. People had shifted their concept of food as a matter for survival and health to one of pleasure and an indication of social status. Changes in the way we shop and the increased use of labour-saving devices meant a shift towards convenience foods. Processed, preserved and packaged foods now account for 70% of all food consumed in Northern Europe. Food processing techniques also increase the amount of salt and sugar 'hidden' in foods.

Nutritional aspects of food choice may be considered important by some members of the population; those who are 'health-conscious' or those with a food-related illness such as diabetes, hypertension or coeliacs. For many people, though, food choice is governed much more by other factors.

Question 1

Which sections of the population in Britain do you think are likely to make food choices based on nutritional influences? Why do you think this is?

Economic aspects of food choice

We tend to think of lack of access to food due to poverty as purely a problem for the developing world, but there is increasing evidence which shows that the very poor in the developed world cannot afford to buy the constituents for what the rest of the population consider to be a healthy diet.

> 'Low-income household – a category which includes all 10 million people on income support in Britain – are far less likely to eat nutritious food such as fruit, vegetables, fish and wholemeal products, and are far more likely to suffer illnesses linked to poor diet' (TES, 2000 p.21)

One report (Walker et.al. 1995) looked at how 48 families with children living on income support managed to eat on a low income. The families ate food that was filling, satisfying and appealing to children, i.e. high in sugar and fat, as they could not afford to waste food. Table 1 shows the cost of 100 kilocalories obtained from different foods.

Table 1 The approximate cost of 100 kcal in different foods, estimated in 1993

Food item	cost p/100 kcal	Food item	cost p/100 kcal
biscuits (custard creams)	2	sausages (pork)	10
sliced bread (white)	3	meat pie	11
rolls (wholemeal)	4	pork (lean)	33
chips (frozen)	4	fish fingers	13
potatoes (boiled)	7	cod fillet (frozen)	95
carrots	20	chocolate bar	8
broccoli	74	corn snacks	12
lettuce	76	apples	19
tomatoes	80	oranges	30
celery	103	milk (full fat)	7
		milk (skimmed)	13

Clearly, the cheapest 100 kcal come from filling foods that are also high in sugar and saturated fat.

The families taking part in the study shopped once a week or more frequently, in local shops, relying on frozen food and convenience meals for ease of storage and to avoid waste. Fresh fruit, vegetables and lean meat were items that they could rarely afford, although all the mothers knew what made up a healthy diet. A group of 197 mothers from all social classes was then asked to agree on what a week's healthy diet for children should be. The cost of this food and the other basic weekly expenditures were then calculated and compared to the families' weekly money from income support. The calculations showed that the cost of a minimum healthy diet cannot be afforded without affecting some other aspect of this tight budget.

Activity 4

Calculate your weekly expenditure on food, remembering to include all snacks and food bought away from home as well as the shopping bill.

What percentage of your income is this?

How would your food intake be affected if your budget decreased? Or increased?

A survey by the National Children's Homes charity in 1991 found that, in the month before the survey, 20% of parents and 10% of children in the poorest families had gone hungry because they did not have sufficient money to buy food. Many mothers did not eat meals themselves, to ensure that their children did not go hungry. Another survey by the same charity found that many young people in poor families had eaten only one meal or no meals at all in the previous 24 hours, and that 90% had not eaten any fresh fruit during the previous day.

Other research not yet published also suggests that nutrition among poor families is worse than ever (TES, 2000).

Even where the situation is not so severe, food choice may be affected by economic constraints, for example choosing sliced white bread because it is the cheapest or choosing unbranded goods, irrespective of quality, because they are cheaper. Research is also being carried out to discover the link between low-income families and accessibility to food (TES 2000).

Question 2

What are the influences on diet of economic restraints?

Social and cultural aspects of food choice

Food plays a major role in the social and cultural lives of people. When talking about how economic constraints affect their lives many poor families mention such social aspects, for example their inability to invite friends over for meals or to visit restaurants. All life events, such as births, marriages and even death, are accompanied by the sharing of food with others. It is also possible that special social events involve special foods, for example expensive delicacies or specially-made cakes.

Religions make use of the symbolism of sharing food in many of their ceremonies, such as the Christian act of sharing blessed wine and bread or the Friday night meal in the Jewish tradition. Abstaining from food (fasting) is also a powerful sign of religious devotion, for example in the Moslem fasting period of Ramadhan. Table 2 opposite shows some of the dietary restriction prescribed by certain world religions.

Food is also linked with various emotions and some people overeat when they are unhappy, depressed or wanting comfort. The enjoyment of food can also have positive effects on our well-being. Humans eat to survive but eating also fulfils social, psychological and cultural needs.

Social changes may also have an impact on food choice, for example with increasing numbers of women working full-time there has been an increase in the sales of convenience foods and ready-made meals. The current social changes, with increasing numbers of single-person households, is likely to eventually show itself in a greater availability of smaller portion ready-made meals.

Activity 5

Can you think of any changes in your own life that have had an influence on your food intake, for example leaving home, long working-hours, having children? Think through in detail what these life changes were and how they changed your food intake. Do you think that they improved your diet or not?

Question 3

What do you think are the different social influences that affect the food choices of young children, adolescents, women and men?

Table 2 Examples of religious dietary restrictions

Food restrictions	
Judaism	eat only animals with cloven hooves and which chew the cud, i.e. cattle, sheep, goats, deer eat only forequarters of cattle and sheep eat only fish with scales and fins no blood, no pork, no game do not mix meat and milk
Islam	no blood no pork no alcohol
Sikhism	no beef
Hinduism	no beef
Days of the year	
Christianity	no meat on Fridays during Lent (Roman Catholics) fast on Wednesdays and Fridays (Greek Orthodox) (with the exception of two weeks) no food preparation on Sabbath (Mormons, Seventh Day Adventists)
Judaism	no food preparation on Sabbath
Time of day	
Islam	during Ramadhan food may not be eaten between sunrise and sunset
Buddishm	monks do not eat after midday
Preparation of food	
Judaism	ritual slaughtering of animals use of separate utensils for meat and dairy products
Islam	ritual animal slaughter
Hinduism	ritual bathing and donning of clean clothes by Brahmins before eating
Fasts	
Christian	40-day Lent fast before Easter and 40-day Advent fast (Greek Orthodox)
Islam	month of Ramadhan 13th, 14th, 15th of each month

Technological aspects of food choice

From the discovery of how to make margarine as a butter substitute to the health-promoting foods now available, technology has contributed significantly to the foods available to us. Large leaps forward were made with the development of the canning process and then the production of frozen foods. The technological developments are too numerous to detail, but you only have to look on a supermarket shelf to see their effects – chocolate flavoured products, meat substitutes, foods with added flavourings, colourings, vitamins and minerals. In the greengrocers, too, technology is at work in the coatings put on fresh fruit to protect them and to prolong their life.

The effect that this has had on food choice has been to make more foods available to us. Freezing and canning mean that we can have fruits and vegetables all year round, as well as foods from other countries. Technological advances have given us 'new' foods, such as Quorn as a meat substitute. Food product developments continue to create new combinations, for example the ever-increasing variety of ready-made soups; pizza combinations and ready-made meals of all kinds.

Activity 6

Look in your own food cupboards and see (i) how much of the food is 'natural', (such as fresh fruit and vegetables), and (ii) how much of it relies on technology for its production, or ingredients, (such as bread, canned food, biscuits).

List what the 'technology dependent' foods are and in what way they rely on technology.

Question 4

How has technology changed the traditional British diet?

The use of genetically-modified foods is a current concern. The debate continues over the safety of these; consumers have probably become more aware and informed over this issue than any other food technology issue. The final decisions, though, will affect what is available, how consumers react to these foods and the choices they make about them.

Technological improvements in the storage and transport of foods means that we can now buy food from almost anywhere in the world, and at any time of the year.

Technological advances in food preparation also affect food choice, for example the widespread use of microwave ovens led to the production of microwave-ready meals.

Question 5

Can you think of another recent technological advance which may have an impact on food choice?

Sensory aspects of food choice

All of the above have considered factors external to the body, but food choice often depends on what we see, smell, taste and feel. Personal likes and dislikes involve these factors but also involve interactions with the state of the body, even a highly-liked food will be less attractive immediately after a big meal.

There exists a preference amongst large sections of the population for sweet-tasting food. Even after achieving a state of 'satiety' on a main course many are able to squeeze in something sweet for dessert. There is an evolutionary explanation for this, as in nature sweetness is usually indicative of energy-rich carbohydrates. It is believed that we evolved in an environment in which there was a scarcity of such nutrients and so developed a preference for them. Also, breast milk is naturally sweet and it may be that a preference for sweetness is developed from birth. Today, however, in Western society there is an over-abundance of sweet foods and a consequent over-consumption.

As with sweet-tasting foods, there is reason to believe that in our evolutionary history salt was often in scarce supply. As the body loses salt in sweat and urine, when too much is lost there may be a craving for salty foods. Nowadays, though, like sugar, there is an abundance of salty foods and consumption often exceeds need.

Sour and bitter-tasting foods, in contrast, are often rejected (usually by being spat out!). This, it is presumed, serves to protect the body against toxic substances.

Exposure to a particular food tends, in the long-term, to increase the liking for that food (Logue 1991). This can be linked to bodily reactions to the food, familiarity or linked situations or emotions. There is some evidence that likes and dislikes can be created or changed in young children by the psychological attitude of the food giver. A pleasant attitude with associated friendly facial expression can increase liking for foods given in that context. Also, if the giver is liked and is also eating that food, there may be an association made, for example of wanting to please or of wanting to be like.

Activity 7

List your own personal likes and dislikes for food items. How many of these are sweet? How many are salty?

Can you say why you have these preferences, what caused them?

Choosing food, then, is not a simple matter. It depends on your knowledge, your location, your situation and internal psychological factors which effect your sensory responses to food.

Food product design and development

Food product design involves developing ideas for products that meet human needs and wants, for example to meet nutritional needs, to combat hunger or to satisfy the desire for a particular sensory experience. It also, of course, involves developing products that will sell and provide a profit for the manufacturer.

Product development teams in the food industry constantly research and develop ideas for new products. Only a few of these ideas, though, are actually new in the sense of being completely original, many food products introduced as 'new' are actually re-worked variations, imitations or reformulations of existing products.

Following research into what consumers want or need, development teams get to work surveying existing products, evaluating their relative success and considering developments that are based on current food products. Teams engage in a series of processes and are led by a Product Development Manager, who has responsibility for the whole project. These research and development processes take time and cost money so manufacturers are keen to ensure that they get the product right first time.

The industrial design practices shown in Table 1 are all part of the general process of food product design and development.

These processes are undertaken by teams of specialists, sometimes working individually and sometimes collaboratively. Table 2 shows the members of the product development team and their roles in the processes.

New product design and development requires the strategic planning of time, resources and expertise. One planning tool that manufacturers use to plan is a GANNT chart, which shows how stages in the overall process will be managed against time. An example of a GANNT chart is shown in Figure 1.

The time taken to develop a new product varies depending on the level of development work needed, extensions to an existing range or slight variations and improvements have a shorter development time than a new concept or a product requiring new technology or production processes.

Table 1

Industrial design practices	Involves
Market and consumer research	Finding out and taking account of user needs and ways of meeting them. Research may be qualitative, for example focus group discussions and taste panels, or quantitative, for example large scale surveys. Gathering, handling, analysing and interpreting data (costs, nutrition, safety and hygiene, statistical data) in order to inform the development of ideas.
Generating ideas	Developing ideas for new or improved products.
Developing design ideas	Methods such as discussion, attribute analysis, disassembly, modifying existing ideas, evaluation of existing products, knowledge of materials and methods, modelling ingredients, CAD/CAM and the use of graphics packages.
Producing a specification	Describing design criteria, a specification describes: what the product has to do and what it must be like the intended user or market other requirements that need to be met, such as aesthetics and costs. Specifications, for the product, and its manufacture, will be produced at the raw material sourcing stage, and are used as part of the quality assurance process.
Evaluation	The use of criteria to judge and test products, for example in relation to performance, meeting user needs, the suitability of ingredients, aesthetics. Evaluating possible production methods and systems.
Trialling and prototyping	Organoleptic (sensory analysis) testing. Considering implications for manufacture and planning for these, for example in relation to product engineering, setting up systems and planning for their control, sourcing ingredients.
Consumer testing	Assessing consumer responses to the product under development.
Pilot	Carrying out a trial run of the product in a pilot plant to assess whether it is feasible to produce the product in volume, to the required standard.
Quality assurance and control	Ensuring quality against design and production specifications, quality assurance.
Marketing and promotion	Initial launch, advertising and promotion throughout a product's life cycle.

Table 2

Role	**Responsibilities**
Product Development Manager	This person has responsibility for the whole process of product development. The job entails liaising between team members and, where opinions differ, making final decisions. She/he may also identify the best sources of ingredients and materials, looking for the best quality at the cheapest price. The PDM decides who will manufacture the product and produces full and detailed product and manufacturing specifications, outlining all the details of the new product.
Product Evaluation Officer (or Product Evaluator)	This person is often a food expert, either a home economist or chef, and is in charge of a sub-team of workers who are employed to formulate and test products. Sensory analysis forms a significant part of the job. Carefully worked descriptors are used to indicate the exact qualities required for the product in order that it could be replicated. Recipes are modified and re-tested constantly. The Evaluator works closely with the Buyer and also has responsibility for the information that goes onto the packaging.
Technical Manager/ Food Technologist	This person looks at the functional properties of the proposed ingredients and tests various types of foodstuffs, ratios and balances of ingredients, bearing in mind how these ingredients meet the specification for the product. This job also carries responsibility for safety and hygiene being applied at all times, and for ensuring that production methods work properly to produce a high quality and economically viable product.
Nutritionist/Home Economist	This person usually works with a sub-team to develop and formulate recipes which meet specific nutritional requirements. Companies need to be seen to be responding to current nutritional thinking so current dietary guidelines are considered and consumer choices reviewed. Products are often nutritionally enhanced or given 'added value' by the experienced selection and use of raw ingredients or additives.
Production Engineer	This member of the team is responsible for the design and development of the production schedules for the new product. Sometimes this involves designing and making machinery to perform particular tasks, or securing the purchase or rental of the equipment needed. The engineer also has responsibility for the production of the packaging for the food product.
Buyer	This person has responsibility for making sure that the new products are, and remain, profitable. He/she monitors and reviews the quality of raw materials and communicates with the supplier if there are any problems or concerns. The Buyer also reviews changes in price and looks at alternative suppliers who may be able to offer a better deal.

Table 2 (continued)

Role	Responsibilities
Product Safety Manager/ Microbiologist	In food manufacturing, there are many stages that involve risks so analytical checks are conducted at regular intervals to assess microbial content. Risks cannot be completely removed, but they can be minimised. Three main areas of responsibility are: Approval – making sure the product is safe before it reaches the shelf, this includes checking raw materials, production processes, holding times, preparation, heating and cooling times. Also shelf-life testing, checking products throughout the storage and shelf-life to ensure they remain safe to eat within the date period given. Surveillance – continuous monitoring of all techniques and microbiological checks carried out on products taken from retailers' shelves. Emergency issues – keeping up-to-date with developments that may affect the safety of the products, including how modifications, developments and reformulation affects the nature, preparation and cooking of the products.
Marketing Manager	This member of the team usually leads the Marketing Department and is responsible for reviewing the market for the proposed product and looking at data collected from the public. The department then tries to predict future trends in products so that the ideas generated have the greatest chance of success, and therefore of making a profit. The Marketing Department will also be involved with the packaging design and advertising and promotional material used. The Marketing Manager works closely with the advertisers when products are launched and campaigns planned.
Legal Department	This department is responsible for ensuring that all information provided to consumers is clear and not misleading, that claims made for the product are true and that the product is sufficiently different from competitors products that it cannot be said to be a copy. This is to ensure compliance with legislation relating to the manufacture and sale of food products.
Accountants	Accountants set budgets and monitor the costs of product development. They make forecasts of the costs of manufacture, determine selling price and forecast profitability.

Stages	Month 1	Month 2	Month 3	Month 4
Market research	✔			
Opportunity assessment/viability	✔			
Analysis of research	✔	✔	✔	✔
Ideas generation	✔	✔		
Testing materials		✔	✔	
Modelling ideas		✔	✔	
Marketing planning		✔	✔	✔
Packaging/labelling		✔	✔	
Trial run		✔	✔	
Production planning				✔
Promotional materials				✔
Product launch				✔

Figure 1: GANTT chart

The stages in new product design and development will now be considered.

Market research

Market research involves investigating and developing knowledge of the market and the competition, analysing and predicting trends and identifying consumer needs and wants so that they can be met successfully. Information can be gained from numerous sources, and falls into two main categories: primary and secondary.

Primary sources include:

- surveys, egs. face to face questionnaires, interviews and discussion with specific groups of consumers – may also be by post or email
- testing and experimenting
- information gained from modelling and trying out ideas.

Secondary sources include:

- books, magazines, journals, encyclopædias
- CD-ROMs and the Internet
- reports and information sheets sent out by companies
- statistical information, egs. spreadsheets, databases
- information from specialist companies such as Mintel and Nielson.

Markets are always shifting and changing and are influenced by a range of factors. Market research may be used to find out whether there is a need for the proposed product, or to find out what the market response and uptake is once it has been launched. Manufacturers and retailers need to understand consumer behaviour in order to respond to it effectively.

Activity 1

Statistics on the national consumption of different food groups are collected by the government and published on the website www.defra.gov.uk

Look on the website at the most recent statistics and note any changing trends in consumption. Try to think of reasons for any changes you identify.

Different stages can be identified during the life cycle of any product, and market research takes place at each of them so that companies are fully aware of how the market is behaving at any point in the product's life cycle. Table 3 shows the stages in the life-cycle of a product and the role of market research at each of the stages. The manufacturer's aim is to maintain sales and profits and their share of the market. They need to be able to judge when a market is

Table 3

Product Life-cycle Stage	**Market research activity**
Development	Research and development of the product, packaging and brand.
Launch	Commercial viability is assessed and informative advertising is used. This is the most expensive phase since the cost of research and development is yet to be recouped.
Growth	Position in the market is established and sales increase. Competitors are likely to enter the field leading to the need for persuasive advertising and price reductions.
Maturity	Product sales peak. More competitors move in and the market becomes saturated. Advertising is increased to maintain the market share and strategies are used to maintain and extend the market position.
Decline	Sales fall dramatically. The product is likely to be replaced or removed from the market or efforts may be made to slow down the decline, for example regenerating interest in the product through special offers, promotions (two for the price of one), offering a product in a different size or packaging, relaunching as 'new and improved'.

saturated and identify the opportunities and gaps in a market, this is the main reason why so many new products are developed.

Generating and developing design ideas in food

Product development (designing) involves developing ideas for putting food ingredients together in novel and interesting ways, or using new processes. Thousands of 'new' foods continually appear on the market but very few of these are based on totally new concepts. Product development is mainly about improvement, imitation or variation.

Concept development

Initial ideas for new food products often begin as 'concept' ideas. These arise from a general notion about a new food product and provide a starting point for more in-depth thinking to move vague, early ideas into a more detailed specification from which a product can be developed further.

One way of thinking about concepts is to see them as hooks onto which ideas can be hung, for example thinking about new food products:

- with nutrition in mind
- for a particular group of people
- for a particular time of day
- for particular occasions
- for a particular time of year
- with current trends in mind.

The concept development process typically involves asking questions about what the product will be like and who it is for, considering the attributes required (attribute analysis), brainstorming ideas and creating a mood or image board to provide a visual picture of the concept being developed.

Where do ideas originate?

- team brainstorming
- taking account of competition, for example through competitor shopping
- suppliers who have a detailed knowledge of trends in the market
- specialist consultants
- customer feedback comments

- failure or success of existing products
- improving, updating or re-launching existing products
- ideas in magazines and recipe books
- recipes from famous chefs and restaurants
- fashion trends
- new ingredients and technologies
- information gained from international travel
- in response to health trends and government guidelines
- constraints, for example manufacturing considerations, the need to reduce costs, the loss of the source of ingredients and the need to seek new ones.

Ideas may develop existing products, for example:

- changing the flavour of a product, such as making a product 'more spicy'
- introducing new varieties, as with seasonal soups or 'soup of the month'
- transferring one product type to another, for example chocolate bars becoming ice-cream bars.

Question 1

What are the main influences on contemporary product development?

Initial concept ideas are gradually refined as they are screened for feasibility. A product specification can then be drawn up. This helps those who will be working on further development to visualise the product they are aiming to develop and provides objective criteria against which proposals can be evaluated.

At this stage consideration has to be given to the working characteristics of possible food materials. Food materials are selected for their particular properties, attributes or characteristics, different ingredients serve different functions. The specification should indicate:

- the sensory characteristics desired – flavours, textures, appearance
- the physical properties needed – aeration, shortening, binding
- the nutritional profile required – low fat, vitamin content, high fibre.

Activity 2

Choose two equivalent products, for example savoury flans or sponge cakes, and compare the ingredients used in each. Look for similarities and differences.

Think of a reason why each ingredient has been selected and used.

The selection of materials will depend not only on the function that it is expected to fulfil but also on its cost and how it is to be processed. It will also be affected by the intended target group for the product, is it to be a luxury product or a 'value' one, and whether it is to be fresh, chilled or frozen.

Manufacturers have clearly defined specifications for each ingredient, to ensure that the product remains the same each time a batch is made. Some food materials have fairly steady prices and manufacturers can rely on their supply and cost to be constant. There are times, however, when these costs increase rapidly due to crop failure, poor weather, transportation difficulties or local or national events (such as strikes or civil war). This is particularly so with the sourcing of ingredients from around the world. Some manufacturers now control their own sources for ingredients, for example some grow their own variety of potato for use in the topping of their pies. This helps to reduce costs and ensures that the desired characteristics are achieved consistently.

Product ideas, recipes and formulations will continue be developed and refined until they are considered to meet the specification. Evaluation is an important part of this development process and includes:

- use of IT, for example modelling
- nutritional analysis
- star profiles and attribute analysis
- disassembly
- sensory evaluation – subjective and objective tests.

The role of sensory analysis

Sensory analysis, or sensory evaluation, is used to focus analytically on the characteristics of a product, through scientific measurement. The scientific principles underlying sensory evaluation ensure:

- testing is rigorous and thoroughly planned
- fair test procedures are used

- controlled conditions are used
- results are carefully measured and recorded
- results are thoroughly analysed and evaluated.

Sensory analysis involves measuring human responses to food products in relation to the senses of sight, feel (texture), sound, smell and taste. Humans gather information about food through sensory receptors, these are nerve endings found in large numbers in specific areas called stimulation sites. The most common stimulation sites include those in the eye, the hands (particularly the fingertips), the mouth and the nose. These respond to stimulation from light waves, sound waves, thermal waves and chemical compounds and affect our individual responses to food products.

Sight/appearance

About 80% of our information is gathered by the eye, it takes very little light to activate the sensory receptors and very low levels of light can be detected. The sight of food influences what we expect it to be like, for example, a wilted lettuce tells us that it is past its best and that the quality is poor before we even try it. Foods are often purchased, or not, on the basis of what they look like. This quick visual assessment can provide information on:

- colour
- size
- shape
- consistency
- visual texture, for example a lumpy-looking sauce
- visual flavour, for example a red jelly indicates strawberry not blackberry.

Touch/texture

Cutaneous receptors are present in the skin, tongue, palate, gums and tooth roots. Initial clues about the texture of foods can be gained by sight, but then handling and cutting foods provides us with information. Cutaneous receptors are sensitive to stresses and strains and give us an idea of the hardness or solidity of food. The most sensitive areas can detect very small changes in force, the size of particles, and spot changes very quickly.

Most of our knowledge about texture, however, comes from the process of eating the food (mastication):

- the hard palate of our mouth contains receptors sensitive to touch and coarseness of food
- the cheeks and soft palate of our mouth can detect different textures
- the roots of the teeth contain nerves that are sensitive to pressure applied to the teeth
- the muscles and tendons in the jaw can detect the amount of pressure needed to chew successfully.

Sound

Sound is perceived when vibrations in the air enter the ear, causing the eardrum to vibrate. Tiny bones in the middle ear transmit these vibrations, which cause the fluid of the inner ear to move. This movement is sensed and neural impulses are sent to the brain. The brain interprets this movement as sound.

The sound made by food, during preparation or eating, gives us information about a product, for example crisps, celery, the snap and crackle of breakfast cereals, the fizz of champagne, the sizzle of a steak.

Smell

Smell helps us anticipate taste and can stimulate saliva production in readiness for eating. Olfactory (aroma) receptors are located in the nasal cavity, but only detect gaseous substances that dissolve in the mucous and stimulate the nerve fibres, nerve impulses then pass to the brain and the more volatile the compound, the stronger the smell.

Smell is important because consumers often accept or reject a product after smelling it, before tasting it. There has been less research targeted at classifying odours, but one suggestion gives seven primary smells. These are:

- peppermint
- pungent
- musk
- floral
- putrid
- ethereal
- camphoraceous.

Smells are difficult to classify as they can be combined with other smells to produce new ones (one of the reasons why the same perfume can smell very different on two people). Humans are thought to be able to distinguish between 4,000 and 10,000 different odours. The sense of smell, however, is quickly 'fatigued', this means that although you may notice a smell initially, continuing exposure to that smell makes you acclimatised so that you no longer notice or barely notice it. The sense of smell is quick to recover, if you leave a room and return within a matter of seconds, you will detect the smell again. The longer the period out of the room, the more you will notice the smell when you re-enter.

Taste

Flavour is detected using a combination of smell and taste. The sense of taste, or 'gustation', occurs when the taste receptors on the tongue detect certain molecules in the food. The tongue can only detect flavour if these molecules are dissolved in a liquid. Different parts of the tongue are more sensitive to certain tastes, because they contain more of a particular type of receptor cell, so different tastes are detected by different parts of the tongue. The front right of the tongue tastes sweetness, the back tastes bitterness, the centre tastes acidity and the left front tastes saltiness. Although receptors for taste and smell are located in different places, they are very closely linked, which is why a heavy head cold and blocked nose causes taste to be suppressed.

Taste is important because it helps us detect unpleasant and potentially poisonous compounds so that we can spit the food out rather than swallow it. Taste also stimulates digestive enzymes in readiness for their job in breaking down foods in the mouth or stomach.

Flavour, or taste, results from four basic taste compounds: sweet, bitter, acid (sour) and salt. A fifth compound, called 'umami' is now thought to exist, although it is much more difficult to identify and detect than the other four. 'Umami' means 'savouriness' in Japanese and could be defined as the 'wow' or 'oomph' factor that some foods give.

Substances taste sweeter when they are warm, and red and orange foods taste sweeter than others.

Bitterness usually increases as a food cools down. Most people can detect very small levels of bitterness and, although it is slow to develop, once there a bitter taste tends to linger. Bitter compounds include:

- some amino acids, such as leucine and valine
- inorganic salts, such as potassium iodide

- alkanoids, such as nicotine
- phenolic substances, such as those found in bitter lemon and orange and lemon peel.

The acid, or sour, taste in food increases as the food warms and is caused by the presence of acids such as:

- fruit juice – citric and malic acids
- vinegar – ethanoic acid
- blackberries – isociteic acid.

Question 2

What descriptors could be given to/used by sensory testers to describe each of the following aspects of a food product: appearance; texture (mouthfeel); aroma; taste?

How is sensory analysis used by the food industry?

At the stage of new product development, sensory analysis is used to help ensure consumer acceptability and optimise the products' qualities. Tasters' preferences may be charted against particular attributes being tested. This information can then be used to create a profile, stating the criteria that the product should satisfy, which can be used as a starting point for development. The profile can also be used to evaluate the developed product to test how well it meets the criteria.

Sensory analysis is also used when existing products are being modified or developed. The improvements may incorporate new technology into the production process, alter food ingredients (perhaps to use cheaper or more easily sourced ingredients), eliminate a problem or just to improve on the quality of the product.

In production, items from the production line are tested against pre-set specified criteria to test accuracy and this may include sensory analysis tests.

Post-production and shelf-life tests also occur at various intervals after production in order to discover how time and storage affect eating quality.

Choosing the appropriate sensory test

There are various types of sensory analysis tests used in the food industry, which is used depends on the product and the purpose of the test. Table 4 describes the tests available, and when each one might be used.

Table 4

Test type	Description	Useful for	Examples
Preference Subjective, only for gauging preference, not attributes or characteristics	Provides information about people's likes and dislikes	Comparing similar products Testing for product acceptability Comparing against competitors' products Where information about consumers' feelings or responses are needed	**Pair comparison** – tasters are asked to state a preference between two samples **Ranking** – tasters are asked to rank in order of preference **Hedonic** – samples are scored according to order of liking or disliking **Scoring** – particular attributes are scored on a scale
Discrimination Objective, for quantifying specific attributes or characteristics, not gauging preferences	Enables evaluation of specific attributes or characteristics	Where information is needed about detectable differences	**Pair Comparison** – tasters are asked to compare two samples for a specific attribute/characteristic/quality **Duo-trio** – tasters are asked which of two samples is different to the control **Triangle** – tasters are asked to identify the odd one out of three samples

Conducting sensory evaluation

The sensory testing environment must be neutral and controlled so that tasters' sensory perceptions are not affected, and to avoid bias. For example, the room itself should be free from all noise and odours; the walls and booths should be off-white in colour and made from non-odorous materials and standard lighting, which is shadow-free, should be used. Access to the test room is usually separate from the entry to the sample preparation area. Testers arrive at their allotted time and no discussion is allowed.

Sensory evaluation is usually carried out in testing booths, which may have a computer monitor and keyboard. The tester sits in a booth so that visual and auditory contact between assessors is not possible. The samples are prepared and slid through to the tester via a hatch, which is then shut. Coloured lights may be used to control the appearance of the samples if tasters would be influenced by colour variations.

A computer program, or written response sheet, prompts tasters to give ratings for the characteristics and attributes of each product tested. Opportunity may also be given for tasters to give their own thoughts or comments. At the end of the test, the computer collates all data and looks at the significance of the findings. From this information the Product Evaluator can suggest best products or identify weak aspects of a proposal which is useful information for product development.

Figure 2: Sensory booths

The development process for a product can take several months, during which time, it may go through sensory testing several times. The results are discussed, reviewed and a detailed report produced to inform the development process.

> ***Activity 3***
> Select a manufactured food product, for example a savoury pie or an instant dessert, and carry out appropriate tests to assess its sensory characteristics.
>
> From your results, produce a profile that could be used by the product development team to draw up a specification for developing a similar product.

Specifications

Specifications are drawn up at different stages in the design process. A **raw materials specification** is used to detail the quality of raw materials required, and to check on these when they arrive at the manufacturing plant. A **product specification** is a detailed description of a product, and specifies the criteria against which quality will be assured. As a new product develops from concept to idea to proposal, its details are refined and it becomes possible to specify them. A specification usually includes such details as:

- product description
- functional characteristics
- the intended user or target

- size/weight or volume
- sensory attributes
- processing and manufacturing details
- safety and quality checks
- raw materials/ingredients
- product shelf life
- cost
- packaging requirements.

A specification should accurately define the product and provide details of its exact composition, this helps to guarantee consistency whenever the product is manufactured. It should include tolerance levels for such factors as colour, texture, size, pH and quality.

Activity 4
Using an existing manufactured food product, write a specification for it using the following headings:

Quality of ingredients to be used

Taste

Texture

Appearance

Size

Shape

Total weight of the product

A **manufacturing specification** is drawn up at a later stage once the product has been trialled and is ready for full-scale manufacture.

Modelling and prototyping food products and their manufacture

Once a product has been through the initial development stages further evaluation, testing, trialling and reviews are needed to ensure that the product has the best impact on the consumer and that it fulfils its potential.

Question 3

Why are testing and evaluation critical in the development of new products?

Food products can be modelled in terms of their nutritional properties, ingredients and costs. These can be analysed and amended to find the optimum product formulation, at the best quality and within acceptable costs.

Manufacturing systems and equipment needed will have been considered during the product development process and these have to be tested, including the identification and control of potential food safety hazards. Computer models are available to assist this process, for example MicroModel can predict how micro-organisms may grow in the food and calculate safe shelf-life and storage instructions for the product.

Question 4

What are the different ways that ideas for food products can be imaged and modelled?

Developing ideas on paper, through discussion or using computer programs is useful, but seeing how a product performs when manufactured and evaluated is a crucial part of the development process. Once the product is accepted, the recipe is scaled up and a first production trial run is carried out. This is designed to check that large-scale production is satisfactory and that the quality requirements identified in the specification can be maintained with large-scale production. It also allows for consumer response to the product to be monitored.

Packaging design

Food packaging is an integral part of the product and subject to similar design and development practices as the product itself. The purpose of food packaging is to protect, preserve, prevent tampering and contain the food. The materials used must protect the food during transportation and storage, and must not taint or otherwise affect the food that is being protected. The packaging specification will include the type of materials and processes to be used, for example vacuum packaging. Government Food Regulations state that food packaging must not be hazardous to human health, contribute to the deterioration of the food or result in unacceptable changes to it.

Packaging is also an important means of communicating information about the product to the consumer. The design brief for the packaging is developed, then checked and approved by the new product development, technical and legal teams to ensure that it meets the quality standards of each. The brief is then passed to a design team who will produce artwork, such as photographs, logos and illustrations. These will be subjected to internal reviews and testing by consumer panels before being accepted.

Food manufacturing

Food manufacture is the name given to the activity whereby edible materials and products are made from plants and animals. Some foods are eaten in a raw or unprocessed state, but usually foods have to be processed for human consumption, to make them palatable, safe to eat and extend their shelf life.

Changes in lifestyle, employment and leisure in the UK have largely determined the changing role of food in people's lives. Food is still a basic everyday need, but there is now an increasing demand for 'fast' and 'ready-prepared' foods and industry responds to this. The technology involved in food production in the UK has developed rapidly over the last 100 years, increasing the number of manufactured food products available.

Scales of production

There are hundreds of food manufacturers in Britain, from large conglomerates to individual self-employed producers. Some make products in their millions, others in much smaller numbers. Table 1 describes the scales of production used in food manufacturing and gives examples of products made by each method.

The scale of production used is often determined by the:

- size of orders or number of products needed
- skills and techniques needed to make the product
- shelf life characteristics of the product
- equipment and tools available
- storage facilities required/available
- actual and intended selling price of the product.

Figure 1: assembly line production

Producing work schedules

Whatever the scale of production, time planning and organisation are essential elements of food production. Food

Table 1

Scale of production	Description	Examples
One-off Job or Craft Production (also known as job production or jobbing)	• a 'tailor made' or 'bespoke' product often produced to meet a customer's own individual requirements • often involves only one or very few workers • requires specialist, skilled workers • can take several weeks, or more, to complete • often commands a high price due to the time and effort involved	hand-made celebration cakes
Batch Production	• used when a specific number, e.g. several dozen to several thousand of the product is required • often used to make pre-set numbers of short shelf-life products • products are identical or very similar • made in one production run or batch • requires a team of workers, involved at different stages (more skilled than for continuous-flow production) • factory equipment may be used to produce variations of same product • time can be lost between making batches of different products on same equipment, due to cleaning and hygiene requirements	pies, salads, sandwiches, continental breads, cook-chill products

Table I

Scale of production	Description	Examples
Continuous Process Production	• the product being manufactured is in high demand and very large quantities are required • production usually occurs 24 hours a day, 7 days a week • a highly automated production system, often computer-controlled, requiring few workers • foods pass through a number of sequenced processes with ingredients being added at various points • identical products made • products usually have a long shelf life • high set-up costs but relatively cheap once up and running • machinery needs regular checks and high maintenance	high demand products such as coca cola, potato crisps, frozen vegetables
Assembly Line Production	• for high volume products • products are assembled bit by bit – often conveyed around the factory • large numbers of workers are employed • workers usually need few skills • purpose built machinery is required • initial start up costs expensive but relatively cheap once production starts • computers often control machinery and monitor the production	filled sandwiches, biscuits

ingredients arrive at the factory and are often processed fairly immediately so that minimal storage is required, many products are made within a 24-hour period of the arrival of ingredients so that fresh foods do not deteriorate and wastage is avoided. Food production plant is expensive to build, run and maintain so to make best use of their resources, manufacturers plan the processing of products with accuracy and precision. Efficient use of staff, equipment and the factory helps to maximise profit and contributes towards maintenance of quality standards.

Several different work schedules are used in the manufacture of food products in order to ensure efficient development and production.

GANNT charts are used to plan projects from start to finish and to sequence all tasks to be performed. Tasks are identified and time limits given for each part to be completed, although several tasks may be going on at any one time. An example of a GANNT chart is shown in Figure 2.

TASK	Week 1	Week 2	Week 3	Week 4	Week 5
Market research	✔				
Analysis	✔	✔	✔	✔	✔
Developing ideas	✔	✔			
Formulation testing		✔	✔	✔	
Consumer testing		✔	✔	✔	✔
Prototype reformulation				✔	✔
Testing				✔	✔
Packaging			✔	✔	
Final formulation					✔
Advertising, marketing and promotional work					✔

Figure 2 – an example of a GANNT chart

Critical path analysis (CPA) is another method used to help designers and manufacturers map out the development of a product from concept to final product launch. Like GANNT charts, this planning tool enables the main stages in production to be mapped out with time limits allocated. CPA's are flexible because they give earliest and latest start times for each activity identified.

Flow process charts can be used to detail the full production system, identifying unit operations, quality control, transport or movement and storage. These charts detail each step in the process of making and can identify where controls are to be applied and the procedures to be followed should a problem arise. They are excellent for linking several tasks and systems together and are often produced using computers.

Activity 1
Draw a flow process chart that could be used in a bakery for the batch production of cheese scones.

Just in Time production (JIT) is an increasingly commonly used method that makes use of Information and Communications Technology (ICT) to help plan the ordering of raw foodstuffs, materials and components so that they arrive at the designated factory only when they are actually needed – hence 'just in time'. Manufacturers using this system operate carefully run and monitored stock control systems so that they are aware of their requirements for each food and packaging material, and storage limitations. Mistakes can prove very costly, for example failure to order the correct quantities of the necessary ingredient can delay production or incur storage costs. If operated correctly, manufacturers have no need for large-scale storage facilities on-site.

Question 1

What are the benefits of JIT production for the manufacturer and the consumer?

Food manufacturing processes

The initial processing of raw materials is known as **primary processing**, for example the processing of wheat into flour. This involves the conversion of raw food materials into edible food materials or components. The complexity of the processing depends on the type of food material, but generally primary foods have received relatively little processing.

Secondary processing is the further processing of primary foods into manufactured edible food products or composites, for example the processing of flour into bread. Secondary processing methods include grading, sorting, cleaning, sizing, mixing, combining, heating or cooling, fortifying, conveying and drying.

Any food manufacturing system is made up of a series of inter-related **unit operations**. These usually begin with delivery and storage, then any preparation that is required prior to processing. The processing operations will depend on the product being manufactured, some examples are given in Table 2. After processing products are packaged ready for transit and distribution.

Figure 3: forming and shaping: sheeting

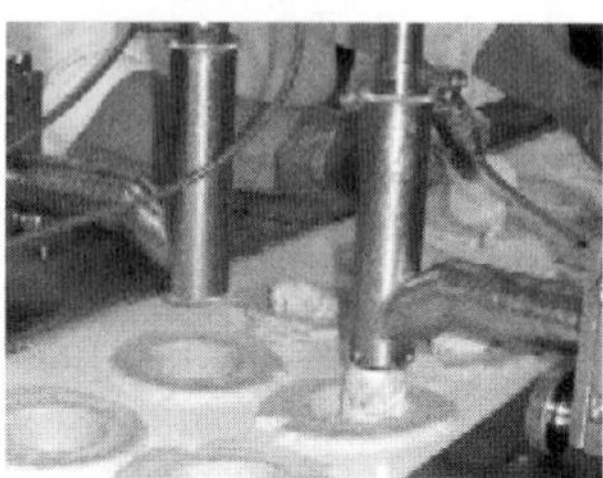

Figure 4: filling and depositing

Question 2

What would be the unit operations for manufacturing a chicken tikka as a chilled food product?

Heating and cooling

The processing of raw materials often involves the heating and/or cooling of products. This can be done to improve or change the organoleptic qualities, to preserve foods for longer by destroying micro-organisms which cause spoilage, or to make them easier to eat and digest, by destroying or slowing down micro-organisms which cause food poisoning. Some materials are too delicate to be treated with heat because their natural characteristics are destroyed or changed unacceptably.

Heating

The methods of bringing about heat exchange are:

- Conduction – heat is transferred through the food by the vibration of molecules at the point of contact with the heat source, this vibration creates heat, which is passed on to the next level of molecules. A good example is frying of foods.
- Radiation – heat is transferred through the air, a specific gas or a vacuum, using electro-magnetic waves. Heat is absorbed by the food but no liquid is involved. Grilling using dry-heat is an example of radiation.
- Convection – as liquids are heated, they rise and expand, allowing currents of heat to move around in the product. Hot liquid rises away from the heat source, being replaced by cooler liquids, which are then also heated. Roasting or baking foods use this type of heat exchange.

Table 2

Process	Description
Extruding • using heat and pressure to shape or form	Mixtures are pumped into a chamber and heated, then forced out through nozzles using pressure that gives a constant stream of mixture. The extruded mixture is then cut into specific sizes. Used for Viennese biscuits, pasta, T.V.P. and savoury snacks.
Forming and shaping • sheeting	As the dough is tipped onto a conveyor belt, it passes through consecutive sets of rollers, each set to a certain thickness that gradually reduces the thickness of the dough. This would be used for biscuits and frozen pastry.
• rotary moulding	Dough passes through a roller with embossed designs on it, this transfers or moulds patterns onto the dough. This is used in biscuit making.
• using individual moulds	These are used for solid products such as chocolate bars or for moulded products that are hollow, such as Cadbury's cream eggs where the two halves of the egg are made separately, cooled then filled with the 'cream'. The moulds are 'glued' together using melted chocolate to form a whole.
Filling and depositing • where a quantity of a mixture fills a specified container	A measured quantity of filling is added to a container, for example pie filling into a pastry case. The quantities are carefully controlled each time.
• where a filling is injected or sandwiched into a product	A product with a gap in it, or two parts of a whole, is filled with a mixture either by injection or spreading. Used for some types of biscuits, buns and savoury snacks.
Enrobing/coating • using vats of a flowing substance	Products to be coated in chocolate pass on a conveyor belt over a vat of melted chocolate that is kept at a certain temperature. The chocolate is raised and floods the belt, coating one side of the product, for example a chocolate biscuit. Products may also be enrobed in other coatings, for example breadcrumbs on fish fingers, batter on vegetable pieces.
• using sprays and nozzles	Products are sprayed from all angles with a liquid substance, sometimes passing through a 'curtain' of the liquid. This is usually done once but for more luxury products, could be repeated. Products are then dried or cooled, with any excess being removed.

Methods of heating used in the food industry include:

- **pasteurisation** – Foods are heated either by batch to 62-63^{0}C for 30-35 minutes, or 72^{0}C for 15 seconds. Both methods are successful and although very small changes in the characteristics of the foodstuffs can be noted, those changes are acceptable. As many micro-organisms are still active within the product, this can only be a short term method of preservation.
- **sterilisation** – Very high temperatures are used, these vary for each product, but milk is heated to 105-110^{0}C for 20-40 minutes. This method changes the appearance and flavour of products, however, and many consumers find this unacceptable.
- **ultra heat treatment (UHT)** – This is a form of sterilisation. Foods are heated to very high temperatures for short periods of time, 130-150^{0}C for 1-3 seconds. These products have long shelf lives – anything up to six months – providing they are stored unopened and the packaging remains air-tight. There is less change to the taste, look or nutritional value of UHT products than those sterilised in other ways.
- **irradiation** – This is a form of sterilisation. Foods with a fairly high water content are exposed to ionising radiation from cobalt 60 or caesium 137. They absorb the rays or beams and chemicals called radicals develop. These destroy many of the micro-organisms and reduce the rate of spoilage/enzyme activity. Many consumers have concerns and fears about irradiation as a process, for example that irradiation masks poor quality foods, causes dangerous chemical changes or becomes radioactive.

Cooling

All methods of cooling are based on the same principles:

- micro-organisms cannot multiply and develop when exposed to very low temperatures
- ice crystals that are formed during the process of freezing make water unavailable to the micro-organisms and preventing their growth
- enzymic activity becomes negligible.

Cooling of food products is usually through chilling or freezing.

Question 3

Why would manufacturers choose to include heating or cooling in the processing operations?

Chilling

This process slows down the rate of reproduction of micro-organisms but does not prevent their growth altogether. Many chilled foods have a shelf life of up to eight days, including the day of manufacture, and they can only be safely stored at temperatures of around 5-8°C. Without these controls, many chilled products would be unsafe to eat. Chilled foods include:

- foods that are eaten cold in their raw state, for example potato salad
- foods that are eaten cold and include raw and already processed ingredients, for example ham salad
- chilled foods that need no reheating, for example cheese or pate
- cooked products that need reheating only, for example pies and pasties
- raw or mostly raw food products that must be cooked before they can be eaten.

The Cook-Chill process

This process is used to produce 'ready-to-heat' foods. After preparation, cooking and portion control, products are rapidly chilled (within 1½ hours) to between 1-3°C. They are stored at low temperatures for very short periods then transferred to refrigerated vans for distribution.

Freezing

In industry, several freezing methods are used, the main ones are:

Blast freezing	Cold air (approximately – 40°C) is passed across batches of food that have been placed on shelves in a cabinet or tunnel until the food freezes. This is particularly successful with unusual and irregular shaped foods, or some that are already packaged, for example some fruits and vegetables.
Immersion freezing	This is a more traditional method where foods are totally immersed for several hours in salt and ice solutions. However, this is time consuming and has been replaced by more efficient methods and processes. Fish can still be frozen in this way using brine.

Plate freezing	Fairly regular shaped foods are prepared in the usual way then placed between refrigerated, hollow, flat metal plates that press on the foods to reduce air. Fish cakes and beef burgers can be frozen in this way.
Scraped heat exchangers	This is used with products containing a higher proportion of liquid. The products are put into a container and freeze on contact with its metal surface. A scraper or paddle scrapes the frozen material from the surface so that unfrozen parts then make contact with the container. This method is used in ice cream manufacture.
Fluidised bed freezing	This is a quick method where fluidised products move along a conveyor belt and jets of cold air are blown up through them. It is used for small vegetables and seafood.
Cryogenic freezing	Foods are sprayed with liquid nitrogen or carbon dioxide, which have extremely low temperatures, and freezing happens instantaneously. Large fans then remove excess gases. The ice crystals formed in the products are very small which means that this method is good for delicate products such as soft fruits.

Preservation

Preservation extends the life of food products for a varying amount of time, depending on the product and the method of preservation used. Preservation works by inhibiting micro-organisms and enzymes so that the rate of spoilage is reduced. Micro-organisms and enzymes can be slowed down by:

- using heat to destroy them
- using reduced temperatures to limit their growth
- using pH levels to inhibit the growth
- removing the water or moisture from the product
- removing air from packaging to create a vacuum, and sometimes adding other gases.

The main methods used to preserve foods are shown in Table 3.

Table 3

Method of preservation	Description
Dehydration	Removing the water inhibits micro-organisms so that they cannot multiply. Products are often reconstituted when used, for example dried soups, herbs, coffee, tea. (see Table 4 for details of the methods of dehydration)
Canning	First developed in 1795, foods placed in bottles, sealed and heated in water were found to last longer, providing the bottles were not opened. Later, cans were introduced to replace the bottles as they were lighter, less easily broken and cheaper to make. Foods are washed and thoroughly sorted before being packed into their containers. Canning sometimes affects the texture, appearance and flavour of foods.
Bottling	The principles used are the same as with canning, but a solution that is either acidic, high in salt, high in sugar or contains alcohol, is added to the product.
Pickling	A highly acidic substance, usually vinegar, is used to cover fruits and vegetables, and a range of spices and herbs is added. The acidity of the substance prevents micro-organisms from developing in the product.
Using salt	This is one of the oldest methods of preservation known. Coating or covering food in a salt solution (brine) reduces the micro-organisms' access to water, so growth is limited. Salt solutions affect the flavour of products and can, therefore, only be used with specific products where a salt content improves the texture or taste of the product.
Using sugar	High concentrations of sugar inhibit the growth of micro-organisms so that they cannot reproduce. This principle is used when fruits are preserved to make jams and marmalades.
Smoking	The smoking process dries the foods. Chemicals may also now be used to replicate the 'smoky' characteristics of foods that have been smoked over wood fires in the traditional way.

Drying foods

A wide range of techniques is used in food manufacturing to remove water from foods, depending on the products being considered. Table 4 describes some of the methods and gives examples of foods processed in each way.

> ***Activity 2***
>
> Select a common manufactured food product, for example vegetables, and survey the variety of ways that the product has been processed, such as frozen, canned.
>
> For each type of product consider the manufacturing processes used and the effect on the food.

Biotechnology

It is often assumed that biotechnology is 'unnatural' human intervention in food production, but it may involve 'natural' biological processes as well as artificial ones. Some biotechnology applications have been utilised for hundreds of years and may be considered to be the earliest examples of food technology.

Some foods naturally contain micro-organisms that pose no risk to humans, provided they are treated or processed correctly. The action of these micro-organisms produces desirable changes in the food that result in characteristic flavours and textures, such as during fermentation.

The most common products produced by biotechnology are:

- cheese
- beer and wine
- bread
- vinegar
- yoghurt
- soy sauce
- pickles
- sliced meats.

The use of the enzyme chymosin, found in calf rennet, has been used in the cheese making process for centuries as a natural biological process. Now the necessary genetic information in the enzyme can be copied so that yeast cells can be used to produce a non-animal derived rennet which is more acceptable to vegetarians, a humane use of biotechnology.

Table 4

Method	Process	Foods dried in this way
Accelerated Freeze Drying (AFD)	Products are frozen and the temperature then increased quickly under a vacuum so that the water held in the ice crystals evaporates to leave a dried product. This process is very quick and has little effect on the quality of the food. Used mostly for luxury products resulting in a high quality.	Coffee Tea Soups
Spray drying	A fine spray of liquid is injected into a chamber where hot air is being blasted upwards from the bottom. As the liquid makes contact with the hot air, the water evaporates and a fine powder is formed. This drops to the bottom of the chamber where it is collected. Used for moist products.	Milk Soup
Fluidised bed drying	This is similar to spray drying. Foods are passed over hot air that is blown upwards through the product. This evaporates the liquid and leaves a more granular effect in liquids and dried small products.	Coffee Lentils
Tunnel drying	Hot air is blown across the product and dries it out quickly. Although the quality of the product remains acceptable, it can shrink slightly.	Vegetables
Roller Drying	The product is usually in a paste form and spread out over rotating drums or rollers. These rollers are heated which evaporates the water.	Instant mashed potato Dried baby foods Sauces
Sun drying	A very traditional method that is slow and only used in hot climates. Foods are exposed to the sun and evaporation of the water slowly occurs.	Tomatoes Dried fruit Fish

Question 4

What are the benefits and drawbacks of biotechnology, from both the manufacturers and consumers viewpoint?

Genetically modified foods

Growers have used selective breeding for years in order to produce a wide range of foods. They choose varieties of plants that have certain desired characteristics or have been specifically developed to exhibit them, for example:

- seedless grapes and oranges
- bright red coloured plums and apples
- stringless runner beans
- high yielding strawberries.

Advances in technology and scientific understanding now allow individual characteristics of plants to be isolated and switched either on or off. For example, some varieties of tomatoes ripen very quickly and their texture softens within a matter of a few days, so that by the time they reach the retailer they could be past their best. Scientists can isolate the gene that causes the ripening and 'switch it off' so that the tomato stays firm for longer (but not indefinitely). Other genetic changes may be used to give:

- greater resistance to disease
- better flavour or colour
- different size
- different or uniform shape
- longer shelf life.

Alternatively, plants may be sprayed with chemicals that make them develop efficiently. These include pesticides to prevent damage from pests such as insects, small animals, slugs; herbicides to prevent weeds from growing; fungicides that prevent moulds growing; and growth promoters that encourage the plant to produce very high yields.

A degree of controversy surrounds the genetic modification of foods, which is viewed by some people as 'tampering with nature'. It may be considered that the long-term effects of biotechnology are not fully known and pose an unacceptable level of risk to the consumer. Due to such concern, many supermarkets have taken the decision not to stock products that include genetically modified ingredients. However, producers are not required to declare this information on food labels. Tracing and guaranteeing the source of ingredients is a complex part of the manufacturer's responsibility to the consumer.

Modern and smart food materials

There are many modern and smart materials now being used in food production. Modern materials are those that are continually being developed through the invention of new or improved processes. Smart materials are sometimes referred to as 'nutraceuticals'. They are called 'smart' because they:

- sense conditions in their environment and change accordingly
- are responsive and appear to think
- may also appear to have a 'memory' and revert back to their original state.

Many naturally occurring food ingredients are smart, that is they respond to heat and light and some changes are reversible, for example gelatin. Additionally, smart materials may be developed as a result of artificially processing materials so that they perform a particular function. Chemically altered or modified starches, for example, respond to differences in temperatures, they swell (thicken) in hot water or when heated, but then return to a flow when cool.

Example of smart food materials include:

- some genetically modified foods
- phytochemicals, for example antioxidants, biflavonoids, carotenoids
- functional foods that claim to improve health in some way, for example Benecol and probiotic products
- modified starches, such as those used in instant desserts that thicken without heating (usually starch requires heating to thicken)
- pizza toppings, where the topping thickens when heated in the oven in order not run off the pizza, but becomes runny on cooling, ready for eating.

Activity 3

Case studies that illustrate smart foods and their applications are shown on the website www.foodforum.org.uk

Look at the section 'F-files' for details of the case studies.

The role of ICT in the food industry

The use of ICT is widespread within the food industry and has revolutionised the way products are developed, tested, manufactured and sold. In particular, Computer Aided Design (CAD) and Computer Aided Manufacture (CAM) have allowed companies to become more efficient, to communicate quickly and effectively and to manufacture on an extensive, global scale.

Computers are useful within the food industry because they can be used to:

- ensure good quality control and repeatability, making production safer for workers and consumers
- increase the speed of production so that a greater volume of products are made
- control complex or continuous processes
- rescue humans from tedious and repetitive tasks
- remove or reduce human error
- reduce the risk of waste and pollution
- reduce maintenance costs and repair, when the system goes wrong, it shuts down before damage can be done to the equipment.

Examples of computer use in food manufacturing are shown in Table 5.

There are also general computer applications which are used in all industries, for example:

Figure 5: Data logging and control

Table 5

Computer application	Use in the food industry
Data handling – spreadsheets and databases	• allows manufacturers to deal with large amounts of information that are needed to organise and run complex production schedules • spreadsheets can be used to record and monitor stock levels and storage quantities of ingredients, calculations can be quickly made for re-ordering stock and costs • databases can be used to access and review a large amount of information such as nutritional composition and recipe files, for research and product development • sensory analysis – programs are used by large retailers to record the results from tasting sessions they conduct with 'members of the public'. These record numerical values awarded for a number of different criteria and the comments and views of those trying out the product. The computer system then collates all data and summarises the findings ready for analysis
Modelling – this is the term given to 'trying things out' before committing to a definite solution, testing and trialling processes and the ways in which materials behave	• costing – often carried out using a spreadsheet program, this allows large scale quantities and prices to be calculated and costs and prices to be modelled • nutritional analysis – using a published database of food values, product developers can try out different ingredients, combinations, ratios, proportions and quantities in order to achieve the required profile for the product • scaling up and scaling down – quantities can be modelled for the different volumes of production, for example batch, continuous or volume production, or changing quantities within one scale of production • safety – programs exist that model bacterial growth in food products, so manufacturers can predict how bacteria will affect the food, when contamination might occur and what factors would encourage it. Manufacturers then use this information to develop ingredients, materials, equipment and processes that will improve the safety of the product. This modelling also helps them to produce storage and user instructions. • HACCP – programs are available to assess potential production risks and highlight the steps that need to be taken to avoid these. • packaging design – graphics software and desktop publishing (DTP) are used to model the shape and overall image for the product. Colours, materials, fonts, pictures and detail for the package can be manipulated until the most noticeable and visually pleasing appearance is reached. Special offers can be highlighted on the package and changed easily for different campaigns and promotions

Table 5 continued

Computer application	**Use in the food industry**
Desk top publishing – this is the term used for designing using drawings, pictures and text with computer software, these allow words and images to be manipulated quickly and easily	• allows manufacturers to test the appearance and impact of the proposed product ideas, packaging, advertising and marketing strategies
Data logging – sensing devices, such as sensors or probes, are used to collect data at regular intervals and to keep a 'log' or record of the information. Deviations from the 'norm' can then be detected using Computer Integrated Manufacture (CIM) or by hand.	data logging can include collecting information on: • the weight (or 'load') of the ingredients • the flow rate – how fast the mixture or ingredients move through the system • the moisture content • the pH levels • the temperature of mixtures, equipment, storage facilities and the manufacturing environment • the pressure levels during manufacturing or during packaging • the speed/efficiency of the entire production system – including machinery, baking, cooling, and packaging
Measurement and control (this often uses data collected from data logging sensors, see above)	• process control – this involves the flow of information passing from the sensors to the decision-maker or controller. Decisions are made about the data gathered by the Production Controller or by computer (using CIM) and adjustments made accordingly. Digital displays can be stored to form a profile of a process or product over a period of time. Processes controlled include temperatures, weights/volumes, size, colour, moisture content, pH levels, speed of flow, handling and use of equipment, microbiological checks • size, portion, tolerance and ratio control – load cell sensors are used to check weight and reject products or materials that do not match the specification or the pre-set tolerances. Computers are used to release specific quantities of an ingredient from large storage silos and to measure the quantities of what remains in the container. This helps with stock re-ordering and helps to ensure that the materials are ordered in the quantities needed and only when they are needed. • colour control – reflectometers check colours by shining a beam of light on the product and monitoring the reflections, different colours and intensities give different reflections. Reflectometers allow the colour of a large number of products to be assessed more quickly and efficiently than can be done with the human eye. • quality control – quality control tests are commonly used to ensure a consistently high-quality product is made, they vary according to the product being made, and include:

Table 5 continued

Computer application	Use in the food industry
	• find and check the critical time control for the setting of gels` • the development of gluten and the production of CO2 during the manufacture of bread and bread products • bake products to the correct size, weight, texture, colour and shape • find and check the critical pH control (acidity levels) in yoghurt and milk products • find and check the critical temperature and proportion control of jams and marmalades, melting of chocolate, rising of bread, gelatinisation of starch mixtures, the storage of foods using freezers and refrigerators, the cook-chill process and the safe re-heating of foods

- word processing for letters, reports, invoices, etc.
- the internet, e-mail and e-commerce to communicate rapidly across the globe. E-commerce allows the ordering of materials through computers. The internet provides up-to-date information for all aspects of research, development and manufacturing and many manufacturers develop their own websites for use by customers
- ISDN and EDI allow computers to communicate directly and enable data to be transferred quickly and cheaply via the telephone link to suppliers, other manufacturers, retail customers etc.
- EPOS to collect and collate data by computer and calculate stock levels automatically
- CIM to produce efficient and rapid manufacture. ICT and PDM systems are used to organise and communicate information between product development teams and the chosen manufacturers
- global manufacturing where products may be researched, designed and manufactured in different locations. Products can be designed in one country, manufactured in another and then sold somewhere completely different with full communication between all parts of the food product development chain
- presentation software to develop presentations for clients and others
- digital imaging to take 'photographs' stored on disk or computer, these can be transferred via ISDN and can also be manipulated to give different visual appearances. Useful for packaging and advertising materials.

Activity 4

Most manufacturers and retailers have their own website for example:

www.kelloggs.com
www.dolmio.com
www.divinechocolate.com
www.unilever.com
www.tesco.co.uk
www.waitrose.co.uk
www.sainsburys.co.uk

Look at two or three of these websites, or others of your choice, and critically evaluate them in terms of the information they present, and how it is presented.

Quality assurance and quality control techniques

Quality Assurance (QA) and Quality Control (QC) systems are used throughout production processes to ensure that products are of an acceptable standard. Quality assurance involves constant monitoring of production to ensure the highest standards are applied throughout. Quality control involves specific checks at particular stages of production. Food companies may use a Quality Management System (QMS) or Total Quality Management System (TQM). This requires manufacturers to provide documents and keep records showing how they have used specifications, quality records, procedures and work guidance instructions in order to provide the best possible service or product.

Quality is a term used to define the qualities, characteristics or attributes of a product, this does not necessarily mean good or high quality, although often the word quality is used in this way. In fact a quality product is one that is consistent, can be replicated time and time again, and does not vary in quality from the accepted standard. Quality food products also need to be safe to eat.

Safety is a key issue when producing foods, if food materials or products are not prepared, handled or treated correctly they can cause food poisoning and illness. Careful control and monitoring of food processing is needed to ensure that foods are of good quality and safe to eat.

There are many things that can contaminate foods, including:

- micro-organisms and enzymes found naturally within the food that cause spoilage and decay

- micro-organisms introduced from external sources, such as poor handling and hygiene of foods
- chemicals that are added during growth or production and those used to clean the production line/factory, some of these could be toxic
- contamination by foreign bodies, including metal, glass and other materials.

Assessing the food risks

Foods can be categorised into one of three types:

high risk foods or products (these are more likely to contain the pathogenic bacteria that cause food poisoning, bacteria find it easy to grow on these foods)	**medium/moderate risk foods or products**	**low risk foods or products**
• High protein products – containing meat, fish, egg, vegetable, cereal, and/or dairy ingredients, or any substitute for these, that need to be refrigerated • Products that are neutral or alkaline (pH 4.6 or above) that have been sterilised in hermetically sealed containers • Infant formula milk	• Dried or frozen products containing fish, meat, egg, vegetable or cereal and/or dairy ingredients or substitutes • Ready to eat sandwiches and meat pies • Fat-based products, such as chocolate, margarine or medium fat spreads, salad dressings	• Acidic products (pH 4.6 and below) such as pickles, fruits, fruit juices and acidic beverages • Raw and unprocessed vegetables • High sugar products such as jam and other conserves • Confectionery products largely made from sugar such as toffees • Edible oils or fats

Food manufacturing sites may be split into high risk and low risk areas. A high risk area is one where high risk foods or cooked components are being handled or where raw and cooked components are mixed together, such as in a pizza or sandwich. Unless particular control measures are taken, it is possible that the food product being manufactured can become contaminated. The high risk areas are separated from other areas, and anyone moving from a low to a high risk area must change their protective clothing and re-wash and sanitise their hands. Foods going into each area are also kept separate.

How manufacturers control systems

For quality to be maintained, food manufacturers need their products to be manufactured in the same way with every production run, whatever the scale of production from jobbing to mass production. This aspect of quality is 'consistency' and is important because:

- Customers come to know the product and purchase it with certain knowledge of what it will look and taste like, and its portion size. Customers then continue to buy the product, this is known as 'brand loyalty'
- Reliability of the product gives the customer confidence in the manufacturing company, or the retailer, and makes them more likely to try other products from the same source
- Products always meet any required legal standards
- Costing the manufacture, and setting the selling price, remain reasonably constant.

Manufacturers ensure consistency by using standard components, for example buying in ready-made pastry to make Cornish pasties, and ensuring component suppliers meet agreed specifications and quality standards. They also use control systems to continually monitor the manufacturing process.

During production, processes are controlled using sensors that are linked to a computer system, this is known as **process control**. The sensors are used to measure and monitor what is happening during the process or sub-system concerned. There are many different types of sensor, each of which is designed to measure only one aspect of the process or product. Sensors can be used to detect:

- temperature – incorrect temperatures can not only make food products unsafe, but can mean that the functional properties of ingredients are not as required, for example when chocolate is melted it has to be kept to a precise temperature to keep it molten, glossy and shiny. If the temperature becomes too low or too high, the chocolate may become unusable.
- colour – using reflectometers that can detect changes in colour during processing and can be used to locate any products that vary from the set acceptable colour range for the product, for example frozen peas.
- weight – using load cells to check the weight of ingredients and mixtures. When attached to storage containers, they can detect how much of an ingredient is stored within that vat.

- moisture content
- pH levels
- flow rate – sensors check the speed of production, or flow, for example of batter for coating products
- metal detectors – these are used to locate unwanted metallic objects which have found their way into products or systems.

What do food manufacturers need to control?

- Hygiene and safety – staff must understand about safe and hygienic working practice
- Equipment – must be regularly cleaned, maintained and checked, and there should be proper cleaning routines
- Staff training and development – staff need to be well trained in the use of tools, equipment and manufacturing techniques
- Sourcing of ingredients and other materials – a good reliable supplier of quality products needs to be found and quality assessed. Some inspections are done visually, particularly where they would be too costly or complex to carry out automatically. Also, expert or consumer tasting panels may be used to select good quality raw materials.
- Stock ordering and delivery – must be carefully monitored and controlled to ensure that the required materials and ingredients match the specification and are where they should be, when they should be there
- Storage conditions – must be kept clean and at the right temperatures for the products; must also be free from pests
- Pest control – an efficient waste disposal system needs to be in operation so that pest infestations are not encouraged. Steps must be taken to ensure that the premises remain animal free
- Date stamping/rotation system – 'use by' and 'best before' dates should be carefully monitored so that ingredients are used in strict rotation and when they are at their best quality
- Measuring of ingredients and materials – measuring equipment must be accurate and quantities checked to ensure the product is successful
- Mixing, cutting and combining – ingredients must be prepared to the correct size and shape for the product being manufactured, they must then be thoroughly combined

- Flow rates and speeds of production – the speed of production must be checked to ensure that the machines are distributing and dispensing the correct quantities and products are being cooked properly
- Construction – products must be put together correctly to ensure portion size and appearance are correct
- Cooking/cooling (heat exchange) – products must be heated or cooled to the correct temperatures for the correct amounts of time so that they are safe and have the desired visual and organoleptic qualities
- Chilling – products should be brought within the safe temperature range as quickly as possible
- Packaging – products must be packaged so that they remain hygienic but are also protected and sealed/closed correctly
- Foreign bodies – detectors (including metal) are used to ensure that no unwanted substances or materials have entered the product at any stage
- Sensory analysis – this is carried out to ensure visual acceptability of the product and that the identified criteria from the specification have been met
- Weight/size checks – this ensures that the weight or size of the product falls within agreed limits and tolerances
- pH checks – these are carried out on some foods to establish acidity and alkalinity levels as this may affect the keeping qualities of the product
- Product storage, transport and distribution – to ensure products are kept at correct temperature and in the right conditions whilst being transported to distribution centres or retail outlets
- Shelf life analysis – to test and ensure product quality remains high at the start, middle and end of the lifetime cycle, for example on day one, three and five for chilled products.

Hazard Analysis and Critical Control Points (HACCP)

The Food Safety Act of 1990 and The General Food Hygiene Regulations of 1995, in England, state that all owners and workers involved in the production of foods must devise and operate a safety system called **Hazard Analysis and Critical Control Point**. This requires that a risk assessment is carried out on all tasks in the manufacture of food, and actions devised in advance to remedy identified problems.

RISK
What could go wrong and at what stage? How likely is this risk to actually occur

ASSESSMENT
An assessment of how serious the risk is and what is the precise potential danger?

The onus for controlling the safety and quality of food products lies with the owner, who has to ensure that a management system is developed to assess, check, predict and minimise the risks that occur during food manufacture. Also, new products cannot be launched without 'Product Safety Approval', so the safety of a product will continually be assessed during its development and HACCP is a key part of this process.

The principles of HACCP

There are seven stages of the HACCP process:

HAZARD ANALYSIS
The details of production are put into a flow diagram with the risks identified at each stage of production

IDENTIFYING THE CRITICAL CONTROL POINTS
These are the identified risks that could seriously affect the safety of the product or the consumer

TOLERANCE LEVELS OR LIMITS ARE SET FOR EACH CRITICAL CONTROL POINT
These can be measured or tested. The probability of a risk occurring is calculated and methods of minimising the risk devised

CRITICAL CONTROL POINTS ARE CHECKED, TESTED AND MONITORED CLOSELY
Regular testing is carried out and measurements taken

A SYSTEM OR METHOD FOR CORRECTIVE ACTION IS DEVISED AND IMPLEMENTED WHEN AND IF A FAULT OCCURS

This puts things right when faults occur

RECORDS OR LOGS ARE KEPT OF ALL PROCESSES, TESTS AND CORRECTIVE ACTIONS AS VERIFICATION

Internal and external parties can review these

PERIODICAL 'AUDITS' ARE CARRIED OUT TO ENSURE THAT THE HACCP SYSTEM IS EFFECTIVE AND WORKING WELL

Changes and modifications are carried out on review of the logs

Critical Control Points (CCP's)

A Critical Control Point is a point at which the safety of the product or the consumer could potentially be jeopardised. These are usually associated with:

- Sourcing the products – unreliable suppliers that are not carefully vetted may provide less than adequate ingredients or materials. Careful research and checks need to be done before production commences.
- Temperature – this can a great source of risk during many different points of manufacture such as delivery, storage, preparation, cooking, packaging and distribution. Food probes and temperature sensors form a large part of the monitoring process here.
- Time – this can occur from the delivery stage all the way through to products being cooked or chilled and stored for the correct amount of time. This is often closely linked to temperature.
- Weight/quantity – at the combination or portioning stage the quantity and weights must be controlled. Cooking times are devised for certain quantities of mixture and may be inadequate or too much if the quantity or size of the product is changed.
- Foreign bodies – during processing and towards the end of manufacture, products should be closely checked for unwanted contamination, this may be biological, micro-organisms, or physical, for example from glass, metals, jewellery.

- Hygiene and safety – workers, equipment and the production environment must be keep the highest hygiene standards possible. Workers often hold a Food Hygiene Certificate to expand their knowledge and understanding of the importance of good hygiene standards.

Non-Critical Control Points

These are often just called Control Points because although they affect the product they are unlikely to compromise safety. These control points can include:

- Preparation of ingredients – ensuring materials are prepared as outlined in the specification
- Sensory analysis – to evaluate and modify products
- Nutrition – ensuring that the product meets any claims made and that the packaging is correct
- Cost control – making sure that the product works with the formulation it has, to the budget that has been set.

Activity 5

Select a recipe for small cakes or a pastry product and draw a flow process chart to show the manufacturing process. On the chart, identify the critical control points and state what would be controlled. You may also wish to consider the non-critical control points.

An example, using a flow process chart for pizza, is shown:

PROCESS	CONTROL POINT
Collect ingredients ↓	T – check use by/best before dates
Weigh out ingredients ↓	M – check accurate measurements of ingredients
Sieve flour	Q – visual check for any physical contamination

Key:
T – time check
M – measurement check
Q – quality check

Limits and tolerances

Once the control points have been identified, measurement limits are set for some aspects such as weights, time, temperatures, for example a product could be set to weigh between 60 and 70 grams. Manufacturers often give themselves a little leeway when setting these limits because it is very difficult to produce identical products using food materials as they react in different ways depending on their source, age, storage temperature and environment. A tolerance level is set that accepts slight variations between products, and a product would be acceptable as long as it falls within the pre-set limits, for example on a product weighing 60-70g a tolerance of 5g either side of the limit would be acceptable. This would be shown as: 60-70g ±5g, so the product may actually weigh between 55-75g.

Food legislation

The purpose of food legislation is to protect the rights and safety of the consumer. The main legislation relating to food products is summarised as follows:

Sale of Goods Act 1979

This states that:

- the product should match the description given
- it should be of 'merchantable quality', i.e. it should not be broken or damaged and should work or perform as described.
- products should be 'fit for their purpose'.

The Consumer Protection Act, 1987
The Trades Description Act, 1968

These make it an offence for companies to make misleading statements or claims about their products.

Food Safety Act 1990

This states that:

- it is illegal to add anything harmful (intended or not) to food, and products cannot be watered down or diluted unnecessarily
- ingredients in a product must be listed and any known allergy-causing products used must be stated clearly

- manufacturers must give 'use by' or 'best before' dates on food along with storage instructions
- food labels and advertising must be truthful
- regular samples will be taken by Inspectors for testing and if standards are not met, prosecutions will follow
- hygiene in restaurants, stalls, factories, cafes, shops and mobile shops, such as burger vans, must be controlled.

Food Hygiene Regulations (1995)

These oblige the food industry to carry out safe and hygienic working practices and to manage risk assessment.

Food labels

By law, food labelling should include the following information:

- product name and description
- name and address of manufacturer or retailer
- ingredients list (in descending order of weight)
- size/weight/volume
- 'best before' or 'use by' date
- cooking and/or preparation instructions
- allergen data, for example 'may contain traces of nuts'
- any claims that are to be made, such as 'reduces cholesterol'.

Labels may also contain nutritional information, but this is voluntary. However, if it is given then it must conform to guidelines in respect of presentation and content.

Trade marks, copyright and patents

A manufacturer can protect a design and brand image by registering it so that others cannot copy it.

Legal standards

Foods may be marketed to appeal to the consumer's concern for health and nutrition, for example a food manufacturer may decide to alter recipes during development to create foods that will have a particular appeal, such as 'reduced fat', 'high fibre', 'low salt'. If they do so, they are likely to make a claim about the product on the packaging. However, products manufactured in the UK that make claims about their nutritional content have to meet the UK Guidelines for Nutritional Claims, which are set by the Food Advisory Committee (FAC). For example, for fat this means:

'Reduced fat' products must have 25% less fat than the standard product on which it is based.

'Low fat' means that the product contains no more than 5g of fat per 100g, or per serving (to be lowered to less than 3g per 100g in 2001).

'Fat free' means that the product contains no more than 0.15g of fat per 100g.

Some products must also meet legal standards that determine the quantities or proportions of certain constituent ingredients, for example beef products may have to contain a certain percentage of beef.

The global perspective

This section has addressed food manufacturing in western, developed countries. It is important to note, however, that consumers in developed countries have considerable choice when it comes to food, they generally can eat what they want, whenever they want. This is not the case for around 14% of the world's population who suffer from a general lack of food, or poor variety in their diet. This may be caused by:

- lack of finance – not having enough money to purchase crops, seed, fertiliser, etc
- unsuitable growing conditions – insufficient water, inappropriate temperature, pest infestations, poor quality soil
- poor use of technology – lack of tools and equipment, possibly because of lack of finance or knowledge
- turbulent situations – civil war, actions by other countries such as trade sanctions.

Many of the same modern technologies may be used in these countries as in western ones. Some developing countries also use methods and technologies more appropriate to their context.

The chemical structure of nutrients

Introduction

To have an understanding of food products it is important to have some understanding of basic food materials, what they are constructed from, how they react and why they react as they do.

The molecules that make up living things range in size from the very small, such as water molecules, to the extremely large molecules that are distinctive of living systems. These are called **macromolecules**. This section shows that biological macromolecules, such as polysaccharides, proteins and nucleic acids which make up food products, are made from much smaller molecules, such as sugars, amino acids and nucleotides, which are joined together as very long chains (that is they are **polymers**).

All biological macromolecules and, of course, their small-molecule 'building blocks,' contain carbon – life is carbon-based. Carbon atoms can form numerous, varied molecular structures. They can form long chains, join with many different elements and form ring structures. Very great use is made of just a few elements, the bulk of the human body being composed of carbon, oxygen, hydrogen and nitrogen. Organic molecules built from these few starting elements provide the majority of the molecules necessary for life.

This section will look at the chemical structure of the main nutrients in the human diet.

The structure of protein

An important group of biological macromolecules are proteins. These are an essential component of the human diet and form 12-18% of the total body weight. Muscles in the body are largely made up of protein, hence the perceived requirement of 'body-builders' for a high-protein diet (in fact, a typical Western diet contains more than ample quantities of protein to support significant rates of increase in muscle mass).

Most people will know what fats or sugars typically appear to look like. Proteins, however, are more difficult to identify because they occur in so many different forms. Some proteins turn up in nature as the structural part of tissue, as in muscle, tendons and hair. Others play an important part in cells as the molecules that carry out and control many of the body's functions.

Chemically, proteins are made mainly from the elements carbon, hydrogen, oxygen and nitrogen. Proteins are long-chain polymers, made by joining together smaller compounds called **amino acids** (so called because of the presence of the *amine* – -NH2 and carboxylic *acid* -COOH, groups of atoms in their structure). Figure 1a shows how an amino acid might be constructed. Alanine, shown in figure 1b, is a typical amino acid and shows the amine and acid group of atoms.

(a)

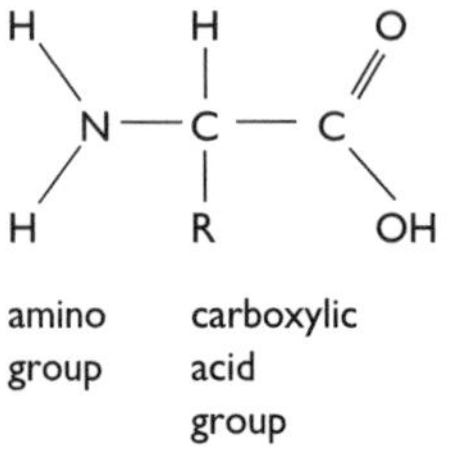

amino group

carboxylic acid group

Figure 1a: an amino acid structure

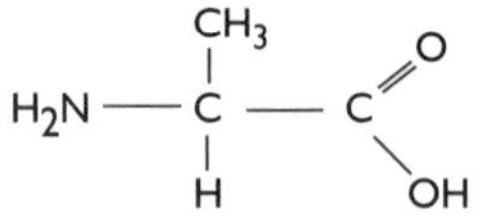

Figure 1b: alanine

There are 20 different amino acids, of which eight are 'essential'; that is, they cannot be made by the body and thus have to be obtained from the diet. There are two others which can only be synthesized from particular essential amino acids, and a further one (histidine) which is made in small amounts, so must also be included in the diet. Arginine is only essential for young children. Table 1 shows the different amino acids, and those which are essential and non-essential.

Question 1

Why are amino acids, which make up proteins, classified as 'essential' and 'non-essential'?

Proteins from animal sources (meat, fish, eggs, milk, cheese) usually contain all the essential amino acids and so can be utilized directly by the body. Those from vegetable sources (peas, beans, lentils, nuts) are usually lacking in one or more of the essential amino acids, so cannot be utilized if eaten alone. For example, cereals contain protein but are lacking in lysine (it is not missing completely, but is present in amounts less than that required by the human body), so eating bread alone would not provide the body with a source of protein. However, if two protein foods, each lacking different amino acids, are eaten at the same meal, then those missing in one food can be supplied by the complementary food, so producing a 'complete protein' which the body can utilize. An example of this would be eating bread, with a shortage of lysine, with beans, which are good sources of this amino acid. These are known as 'complementary' proteins.

Table 1 essential and non-essential amino acids

Essential amino acids	Amino acids synthesized from essential amino acids	Non-essential amino acids
lysine	tyrosine (1)	glycine
methionine	cysteine (2)	alanine
threonine		serine
leucine		proline
isoleucine		glutamic acid
valine		glutamine
phenylalanine		aspartic acid
tryptophan		asparagine
(histidine)		
(arginine – for young children)		
	(1) synthesized from phenylalanine (2) synthesized from methionine	

Question 2

Explain how different protein foods can complement each other, and why this might be useful in some diets.

All the amino acids have the standard type of molecular structure; they vary only in the identity of the so-called R group (shown on figure 1a). The structure of the R group is crucial because it determines both the shape and chemical properties of an amino acid, which, in turn, determines the structure of the amino acid polymers, i.e. proteins.

How, then, is a protein formed from amino acids? Figure 2 shows the chemical reaction that can occur between two of the simplest amino acids, glycine and alanine. After the glycine and alanine have joined together, in

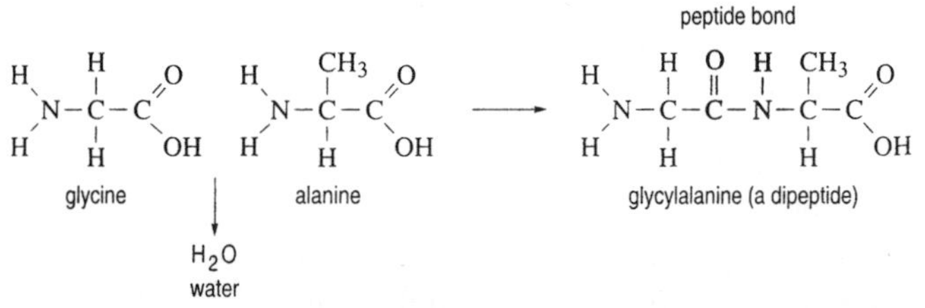

peptide bond

glycine

alanine

H_2O
water

glycylalanine (a dipeptide)

Figure 2 how two amino acids join together to form a dipeptide

what is known as a **peptide bond**, they form a **dipeptide** called glycylalanine. (Notice that the formation of a peptide bond involves the elimination of a molecule of water).

The groups at either end of the dipeptide are free to undergo the same reaction with other amino acids. If fact, in living systems the chain always grows by addition to the free carboxylic acid group (-COOH). The addition of another amino acid would give a tripeptide, and so on. In this way a long chain of amino acids can be built up, i.e. synthesized into what is called a **polypeptide** chain, as shown in Figure 3. The terms 'polypeptide' and 'protein' are sometimes used synonymously. The number of possible combinations of the 20 amino acids in a polypeptide chain is enormous.

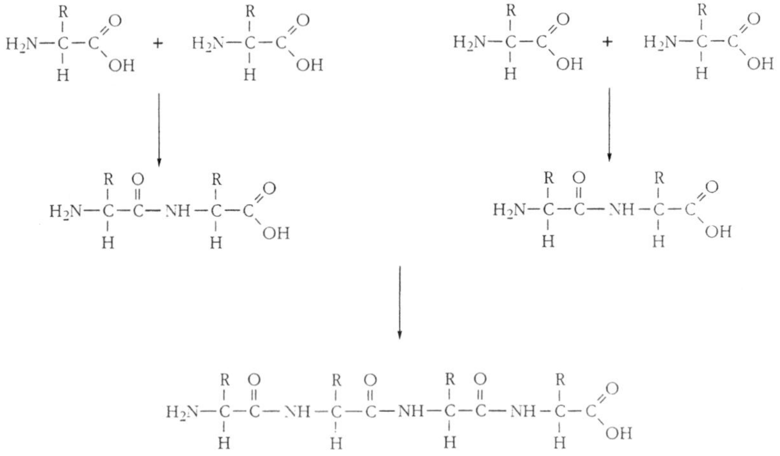

Figure 3: a generalized representation of the formation of a protein from amino acids. Note that in each step a molecule of water is lost

It is this enormous variety of amino acid sequences in proteins that gives their wide range of functions within the human body. Some, like insulin, are hormones, i.e. chemical messengers secreted into the bloodstream. Others make up the overall structure of the human body, for example the muscle proteins actin and myosin, keratin found in skin and hair, and connective tissue proteins such as collagen which fill up the spaces between cells in the body. Other proteins form part of the immune system, defending the body from infection, for example the antibody proteins in the blood. Haemoglobin is a protein found in the blood and transports oxygen to the body cells. Another group of proteins catalyse (i.e. speed up) the many chemical reactions which take place in the body; members of this very important group are called 'enzymes'.

Now, if you imagine a protein molecule surrounded by water molecules, the R groups in the protein that carry an electric charge, negative or positive, will tend to associate with water molecules easily; they are said to be 'hydrophilic' (water-loving). Likewise the polar but uncharged R groups will also be hydrophilic.

On the other hand, the amino acids with non-polar R groups will tend not to associate with water molecules. In fact these so-called 'hydrophobic' (water-hating) groups will tend to cluster together in the interior of the molecule in order to avoid contact with water molecules. The protein chain will thus fold up into a shape that maximizes the interaction of the hydrophilic R groups with the surrounding water molecules and, at the same time, maximizes the interaction of the hydrophobic R groups with each other inside the protein molecule, away from water. This is shown in figure 4.

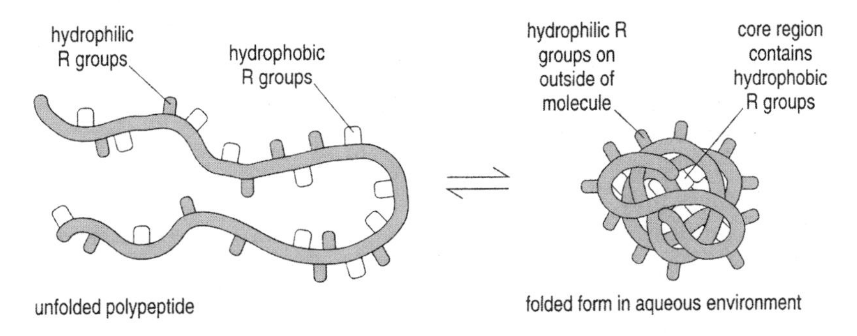

Figure 4: how a protein folds due to the interactions of its hydrophilic amino acid R-groups with water, and its hydrophobic amino acids R-groups with each other

The protein will therefore assume a shape determined by the sequence of its constituent amino acids. Inevitably some charged R groups will find themselves forced to be in the interior of the molecule but pairs of these with opposite charges will attract each other and form ionic bonds within the protein chain which will further stabilize the structure.

Question 3

In what way do amino acids determine the shape of a protein molecule?

Question 4

Identify the R-groups in the amino acids in figure 5 (see page 68).

There are two other types of interaction which are very important in the folding of protein chains. The first is called **hydrogen bonding**. Look back at figure 2, the structure of the peptide bond. You will see that it is made up of a

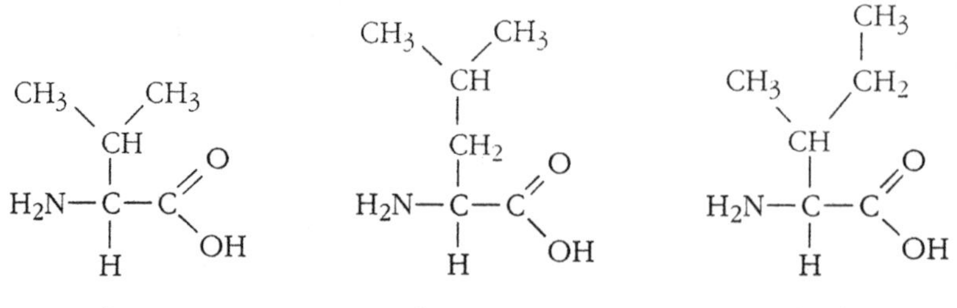

valine leucine isoleucine

Figure 5

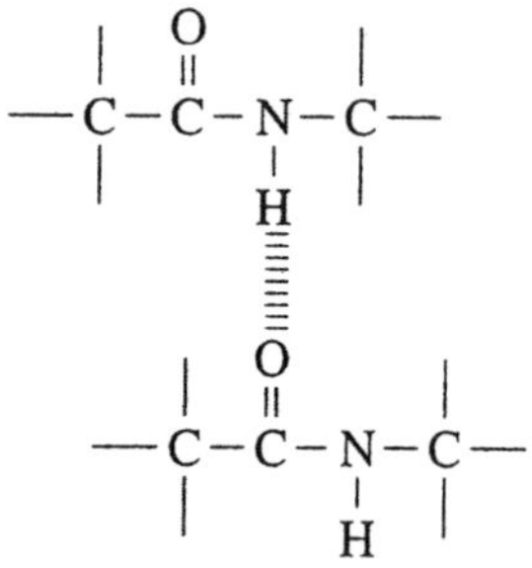

C=O and an N-H group. Oxygen atoms (shown by 'O'), like water molecules, are strongly electronegative. They tend to pull electrons towards themselves, resulting in a slight negative charge at the oxygen atom and a slight positive charge at the carbon atom ('C'). Nitrogen atoms are also electronegative, so there will be an excess of negative charge on the N (nitrogen atom) and a small amount of positive charge on the H (hydrogen atom). This means that a C=O group of one peptide bond adjacent to an N-H group belonging to a peptide bond elsewhere in the chain will have a small but significant mutual attraction, referred to as a 'hydrogen bond'. This is shown in figure 6.

Figure 6: hydrogen bonds, shown as hatched bars

Individually, hydrogen bonds are very much weaker than ionic or covalent bonds but in a protein molecule there will be a large number of these, resulting in a considerable stabilizing effect.

There is one last way in which the shape of a protein can be stabilized and this is by **disulphide bridge** formation. As the name suggests, this involves covalent bond formation between two sulphur atoms of amino acids in different positions along the chain, giving an atomic 'bridge' across the molecule.

Methionine and cysteine are the two common amino acids that contain sulphur atoms. Both can be found occasionally throughout most protein chains, but because of its particular chemistry, it is only the cysteine which take part in disulphide bridge formation. As the chain folds, two cysteine amino acids may come into close contact, allowing the sulphur atoms in their

R groups to react together to form a disulphide bond (bridge). Insulin is a protein made up of two polypeptide chains joined together by two disulphide bridges; in addition, in one of the chains there is an intra-chain bridge too, see figure 7.

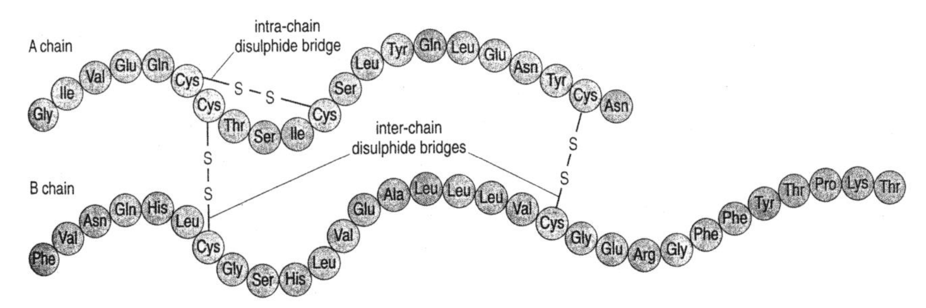

Figure 7 simplified structure of insulin, a protein containing disulphide bridges

Question 5

How do hydrogen atoms and sulphur atoms affect the shape of a protein molecule?

Now that you are familiar with the various forces that maintain the three-dimensional structure of protein molecules, you can think about the hierarchy of protein structure in more detail.

The *sequence* of amino acids unique to a particular protein is called its **primary structure**. Any changes in the primary structure, even changing just one amino acid, can have great implications for the proteins' three-dimensional structure and, therefore, its function.

Question 6

Why would a change of just one amino acid affect the structure of the protein?

Protein chains form two main types of structure: small and globular or long and fibrous. With both types, there are two common chain folding patterns: the alpha-helix (a spiral) and the beta-sheet. In the alpha-helix the turns of the helix are held together by the peptide N-H to C=O hydrogen bonds described earlier. It is similar for the beta-sheet, but here the bonds link adjacent separate chains or parts of chains. Both these types of regular structure are referred to as the **secondary structure** of the protein.

The overall shape unique to a particular polyteptide chain is called its **tertiary structure**. We have already looked at how disulphide bridges can help to stabilize tertiary structure.

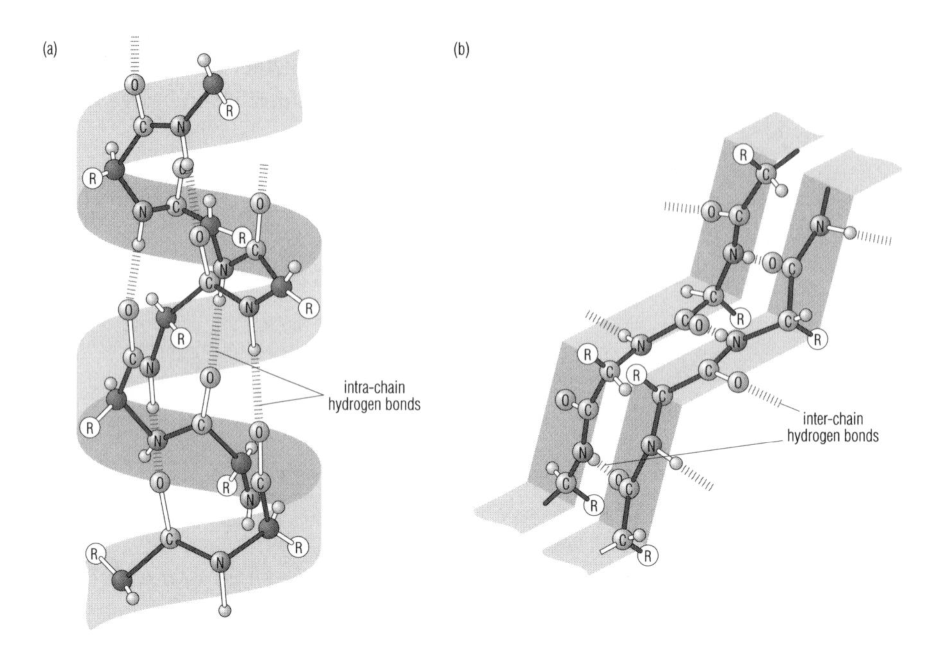

Figure 8 protein secondary structures: a) alpha-helix and b) beta-sheet The very small spheres are hydrogen atoms

Fibrous Proteins

Fibrous proteins are generally used as structural material in the body. They are an important part of skin, hair and fingernails. At the molecular level, they tend to have repeating segments of similar amino acids.

Collagen is the most abundant protein in the body and is a fibrous protein. It is used to form the connective tissue that surrounds the bundles of muscle fibres and connects muscles to the skeleton. Bones consist of a matrix of collagen containing calcium phosphate. The polymer chain is very long, containing about a thousand amino acid residues. The main repeating segments involve the amino acids: glycine, proline and hydroxyproline, although the exact make-up varies slightly from one molecule to the next.

With such long chains of molecules you might think that a small plate of protein would be rather like a plate of cooked spaghetti, with all the chains entwined haphazardly, but this is not the case. The polymer chains group together in threes and wrap around each other, rather like a rope or a plait. The strands in the 'rope' or 'plait' are held together by hydrogen bonds. The strands in the 'ropes' then line up to give a strong tissue. Extra cross-linking via covalent bonds between the 'ropes' gives added strength. Such cross-links

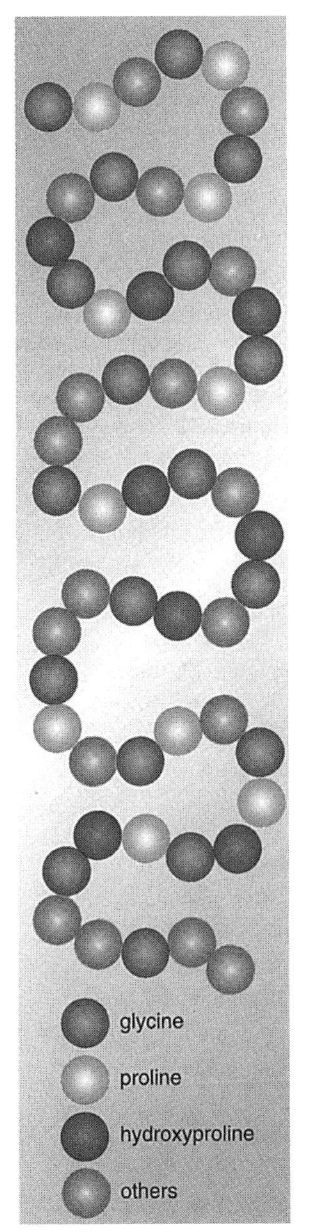

Figure 9 a section of the fibrous protein collagen. Notice that every third residue is a glycine and the sequence – glycine; proline; hydroxyproline recurs frequently.

are most abundant in connective tissue where great strength is needed, for example in the Achilles tendon.

It is a common observation that meat from young animals is more tender than that from older ones. This is because as the animal ages, the number of cross-links between the collagen increases, making it tougher. When meat is cooked in water, molecules of water squeeze in between the collagen chains and replace the hydrogen bonds between the chains with hydrogen bonding to water molecules. This breakdown of the cross-linking means the meat is easier to eat. On cooking some collagen can also dissolve into the water, forming what is known as gelatin.

Meat tenderizing

Meat can be tenderized in several ways before it is cooked by breaking down some of the links between collagen chains. Soaking meat in acid, such as vinegar, destroys some of the collagen on the surface of the meat. Papain from papaya, bromolain from pineapples or ficin from figs are enzymes that break the collagen up; thus meat can be tenderized using the fruit. Alternatively, a commercial extract can be used. Physical pounding or grinding, as used to make hamburgers, also breaks up the collagen. Finally, extended cooking, as in boiling or stewing, also disrupts the collagen. Eventually, meat falls apart!

Figure 10 a) the triple helix structure of collagen
b) longitudinal alignment of collagen 'ropes'

Globular Proteins

Globular proteins are of fixed chain length and have a strict order of amino acids. Thus all the molecules of a particular globular protein are identical, which is not the case with a fibrous protein. Each long-chain polymer molecule of a globular protein has the same specific job to do, be it breakdown of food, transferring information in the body, or fighting bacteria. Only if it has the right sequence of amino acids will it have the correct activity. Although the sequence of amino acids in any particular type of globular protein is specific to that protein, the sequence varies greatly between one type of globular protein and another.

However, a globular protein is characterized not only by its primary structure. As with fibrous proteins, globular proteins have a precise overall shape rather than just a random one. With fibrous proteins we saw how cross-linking between chains is important to produce a strong polymer. With globular proteins we no longer concentrate on cross-links between different chains of polymers but on cross-links *within* a particular chain, leading to the joining of one section of the chain to another. Figure 11 shows how a globular protein is folded into one particular shape.

Figure 11 a 'ribbon' model of ribonuclease. This is a globular protein which consists of a chain of 124 amino acid residues. Sulphur-sulphur cross-links have been marked.

If the cross-links are broken, the chain unfolds and takes up a random structure, and the old structure cannot easily be reformed. This is what happens when you heat an egg, (see box below). The polymer chains are cross-linked so that they have a precise shape in the egg, but when you heat the egg the cross-links are broken and the chain takes up a random shape. In the same way that you cannot unscramble an egg; the random coils of polymer chain will not reorder themselves. Loss of shape of a globular protein is called **denaturation** and is usually accompanied by the loss of the activity of the protein.

Cooking eggs

Most materials turn from solids to liquids to gases on heating, but heating eggs leads to them solidifying! Eggs contain many globular proteins, which fold into compact balls. High temperatures denature these proteins so that the chains tangle with each other and become cross-linked by hydrogen bonds. This leads to a solid three-dimensional network of protein molecules. Heating eggs with milk also leads to a progressively more solid product. Overheating creates too many hydrogen bonds between the proteins, and some of the water is squeezed out. Hence, overcooked quiche becomes runny, custard gets lumpy and scrambled eggs rubbery! Adding salt or lemon juice (acid) causes the proteins to denature and coagulate more easily. So boiling eggs in salt water means that if the eggshell cracks, the protein released quickly coagulates and seals the hole. Egg whites can be denatured by whipping them. This also incorporates air into the mixture. Thus air is trapped in a three-dimensional network of protein leading to a light fluffy form. Baking causes more protein to coagulate, giving a solid meringue. Fats tend to coat the denatured proteins preventing them from coagulating, so it is important not to have egg yolk, which contains fats, in the egg whites.

Question 7

Many proteins foods contain fibrous and globular proteins. What is the different effect of these two types of protein when a food is heated?

The structure of carbohydrate

Carbohydrate in the diet is usually associated with sugar, or perhaps bread, but there are many compounds that come under the heading of carbohydrates. Sugar can be:

- sucrose, which is ordinary sugar obtained from sugar cane or sugar beet
- lactose, from milk
- maltose, from malt
- glucose, grape sugar
- fructose, fruit sugar

In addition there are other carbohydrates: cellulose, starch, pectin, glycogen and gums. The alginates or carrageenans that are obtained from seaweed also belong to this group and are used to make, among other things, synthetic cherries for the bakery trade and 'apple' and 'apricot' pieces that are sold as pie fillings.

Clearly, carbohydrates are an important and widespread class of organic compounds, but even this list does not really do justice to the wide range of different carbohydrates in nature, which utilizes them in a variety of ways such as for structural material and energy stores. Nevertheless, despite the variety, they are all based on the same unit.

The structures of carbohydrates are complex, but don't worry too much about this; the main factors will be highlighted. The simplest of the carbohydrates is **glucose** (sometimes referred to as dextrose – check food labels). This is found in grapes (7% by mass) and onions (2%), but the richest source is honey (31%). It plays an essential part in the metabolism of all plants and animals and is the main product of photosynthesis, the method by which plants store energy from the sun. Human blood has about 80-120mg of glucose in every 100 ml and glucose is the only sugar that plays a significant role in human metabolism.

Glucose has the molecular formula $C_6H_{12}O_6$. Its molecular structure, see figure 12, consists of a ring of six atoms made up of five carbon atoms and one oxygen atom. Attached to the carbon atoms are four OH groups and one CH_2OH group. This structure really needs to be viewed as a three-dimensional one.

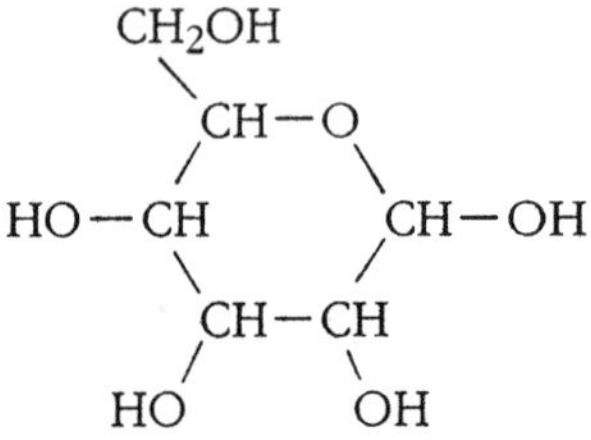

```
   CH2OH
      \
       CH—O
      /     \
HO—CH        CH—OH
      \     /
       CH—CH
      /     \
    HO       OH
```

Figure 12: ring structure of a glucose

Another simple carbohydrate is fructose, which is also found in honey (about 35%). Gram for gram, fructose is about twice as sweet as table sugar, so only half as much is needed to sweeten a meal. Fructose, therefore, can be part of a calorie-controlled diet. It can also be used by people who suffer from the disease diabetes mellitus, since, unlike glucose, it does not require insulin for its use by the body. Fructose has the same molecular formula as glucose, but a slightly different structure. About 75% of fructose has a ring structure of six atoms, just like glucose, however, the CH_2OH group is attached to a different carbon, as shown in figure 13a. The other 25% has a ring structure of five atoms, as shown in figure 13b.

Glucose and fructose, because they only involve one sugar unit, are known as **monosaccharides** (from the Latin for sugar – saccharide).

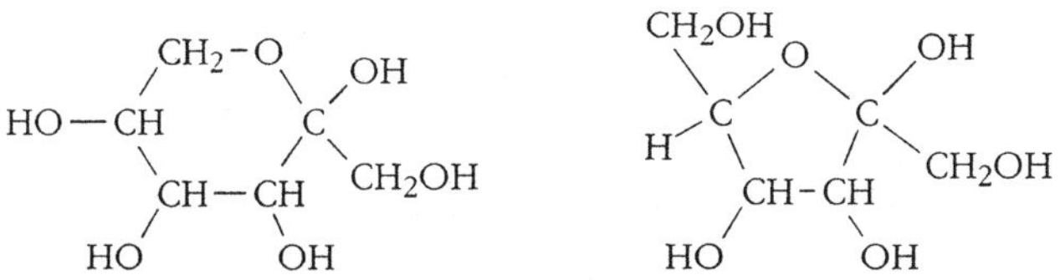

```
      CH2-O                          CH2OH
     /     \   OH                      \     O       OH
HO—CH       C /                         \  /   \   /
     \     / \                       H - C       C
      CH—CH   CH2OH                       \     /  \ CH2OH
     /     \                               CH-CH
   HO       OH                            /     \
                                        HO       OH
```

Figure 13a: simple six-atom ring – structure of fructose
Figure 13b: simple five-atom ring – structure of fructose

The next carbohydrate structure to look at is **sucrose**. This is most familiar as the sugar used to sweeten tea or in cooking. Sucrose obtained from sugar beet is the same as sucrose obtained from sugar cane. About two million tonnes of sucrose are used annually in the UK. Sucrose has the molecular formula $C_{12}H_{22}O_{11}$ and is made by joining a glucose unit and a fructose unit, as shown in figure 14.

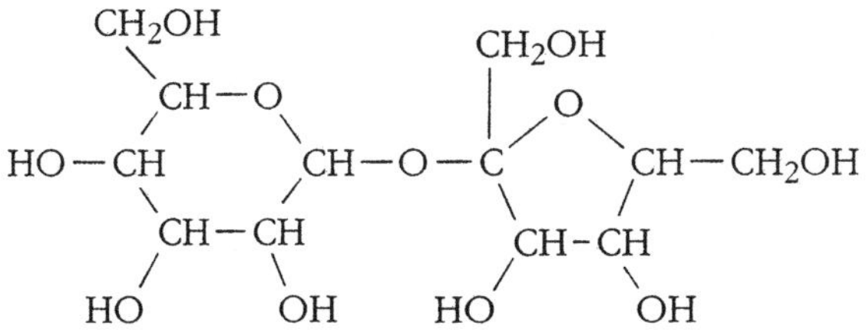

```
     CH2OH
         \                   CH2OH
          CH—O                 |   O
         /     \               |  / \
   HO—CH        CH—O—C           CH—CH2OH
         \     /        \       /
          CH—CH          CH-CH
         /     \        /     \
       HO       OH    HO       OH
```

Figure 14 ring structure of sucrose

Sucrose involves two sugar units, glucose and fructose, so is known as a disaccharide. Other **disaccharides** are:

- maltose, made from 1 glucose unit + 1 glucose unit
- lactose, made from 1 glucose unit + 1 galactose unit (another monosaccharide)

Notice that all the sugars have a name which ends in 'ose'. If you see an ingredient on a food label with this ending you can be fairly sure that it is some form of sugar.

Question 8

Which sugar 'unit' is found in all types of sugar?
When glucose is combined with glucose, what sugar is produced?
When glucose is combined with galactose, what sugar is produced?

The last important class of carbohydrates is the **polysaccharides**, which are polymers (long chains) made from monosaccharides.

Starch is an important polysaccharide in the human diet. It is the chief food reserve of all plants, stored in stems, for example in the sago palm, or in tubers, for example as potatoes and cassava. Under a microscope, starch granules can be seen in the cells of such foodstuffs. These starch granules are made of two components, amylopectin and amylose, in a ratio of about 3 or 4:1, by mass. Amylose is a polymer that contains between 70 and 350 glucose units, joined as shown in figure 15.

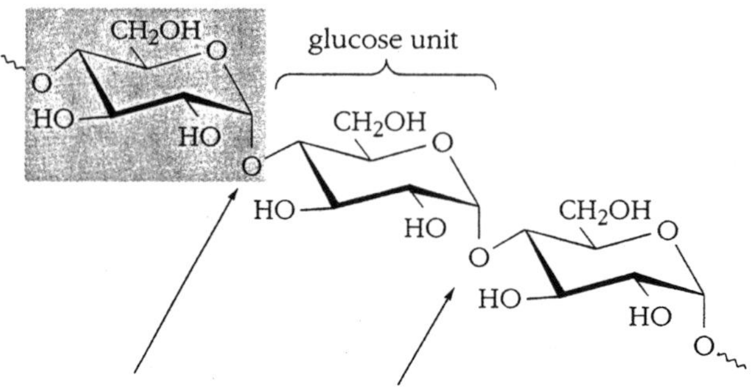

Figure 15 the structure of amylose

This way of joining the glucose units in amylose is known as an alpha-linkage. This linkage means that amylose *can* be digested by mammals, including humans, whereas other polysaccharides cannot, for example cellulose. In digestion, these linkages are broken and the polysaccharide is reduced to disaccharides and then to individual glucose units, which the body can use.

Amylopectin is a much larger polymer, containing up to several million glucose units. A portion of amylopectin is shown in figure 16

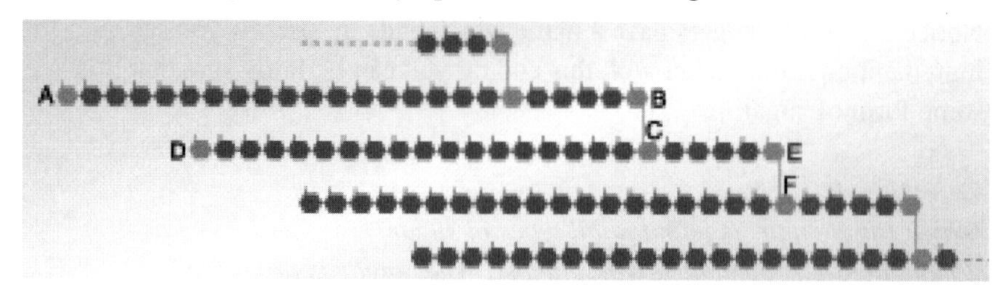

Figure 16 part of an amylopectin molecule

Each hexagon represents a glucose unit. 'AB' and 'DE' are short chains of about 24 glucose units connected at 'C'. Another connection is shown at 'F'. Branching is so extensive that a bush-like structure is formed, as shown in figure 17.

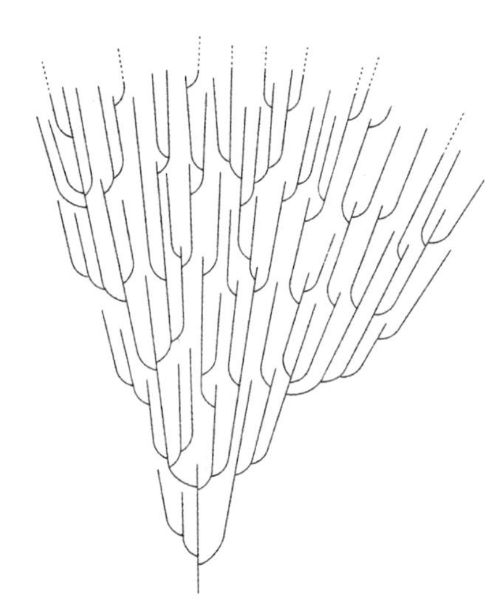

Figure 17 one possible pattern of branching of the chains in amylopectin

Just like proteins, the carbohydrate polymers can cross-link by hydrogen bonding either between straight-chain portions of the amylopectin or between straight-chain portions of amylopection and amylose. Such cross-linking makes starch granules hard.

Cellulose, another polysaccharide, is the main structural carbohydrate of plants and, as such, is widely distributed from the toughest tree trunk to the softest cotton wool. **Cellulose** is the most abundant organic compound on earth. This polymer is made by joining together glucose units, as shown in figure 18, and as many as 12,000 units may be linked together.

oxygen linking glucose units together

Figure 18 glucose units linked to form cellulose

This method of linking allows the cellulose to form very long straight chains, which are maintained by hydrogen bonding between the oxygen atom in one ring and an OH group on an adjacent glucose unit, as shown in figure 19.

Figure 19 hydrogen bonding in cellulose

The long chains in cellulose lie side by side and are cross-linked by hydrogen bonds between the OH groups on the rings in a very ordered way. In plant material this forms small fibres several micrometres long and about 10-20nm in diameter. This kind of molecular architecture gives rise to very strong fibres. Despite being the most abundant natural product, cellulose is indigestible to humans and to most other carnivorous animals. It can, however, be used by ruminants, such as horses, cows and rabbits, which harbour bacteria in their digestive tracts that can breakdown the cellulose.

Cellulose does play an important role in our food as 'dietary fibre'. Even though it is not digested, it passes through us and helps to eliminate waste from the body.

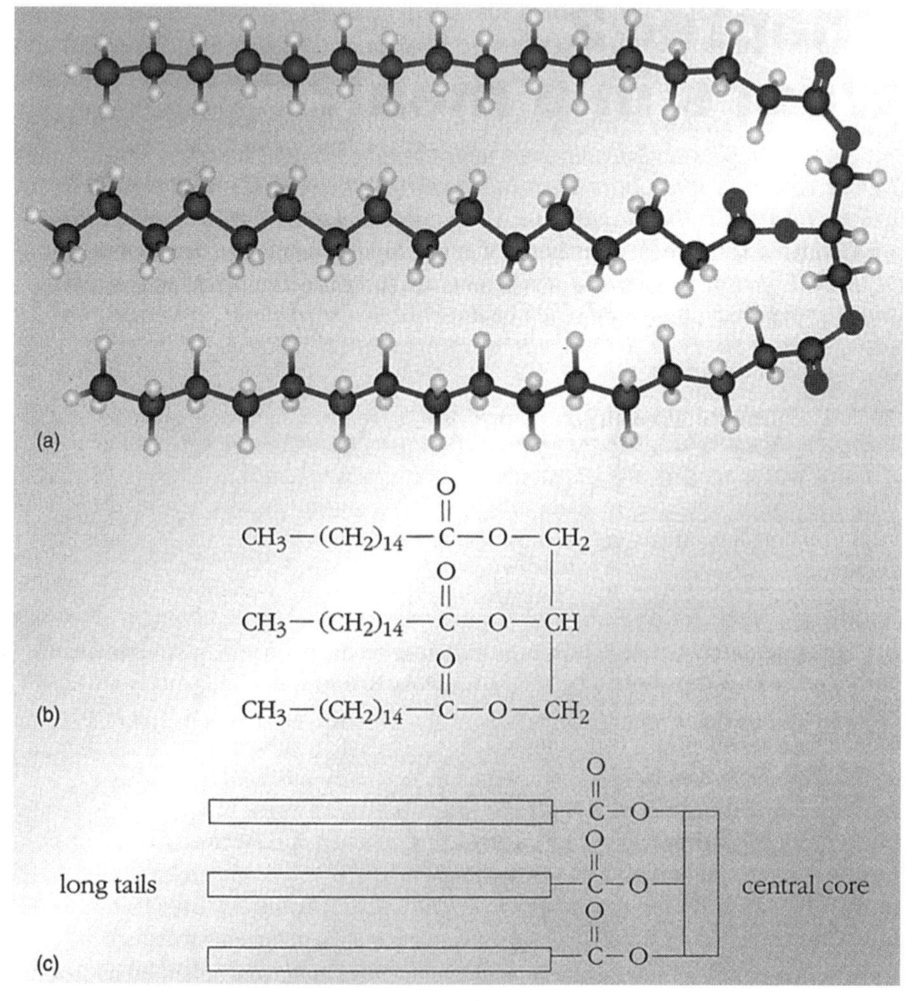

Figure 20 a) ball and stick model of a triacylglycerol molecule
b) structural formula
c) simplified model of a triacylglycerol molecule

Question 9

Why is cellulose not digested by the human body?

The structure of fats and oils

In scientific terms, fats and oils are known as 'lipids', as this embraces a wider variety of substances. However, here I shall continue to refer to fats and oils as this relates more directly to food technology.

In contrast to the proteins and polysaccharides discussed above, although fats and oils form large aggregates, they are not actually polymers. As with carbohydrates, the molecules of fats and oils contain the elements carbon, hydrogen and oxygen. However, lipids contain a smaller proportion of oxygen than do carbohydrates.

The most well-known types of fats and oils are called triacylglycerols (also still called **triglycerides**). These compounds are the main constituents of vegetable oils and animal fats.

Triacylglycerols are the most abundant fats in the body; our capacity to store them is almost unlimited. Body fat serves not just as an efficient energy source but also as insulation (fat deposits under the skin) and as protection (fat deposits around the major organs).

Triacylglycerol molecules consist of three long tails connected to a central core. These long tails are chains of carbon atoms, usually between eight and twenty atoms long. The central core consists of three carbon atoms with their associated hydrogen atoms. Figure 20 (see opposite) shows simple models of triacylglycerol molecules.

Figure 21 (see page 80) shows a more complex model, detailing the structural formula.

As you can see, triacylglycerol is made from three molecules of fatty acid (the 'triacyl') and one molecule of glycerol. The -COOH (carboxyl) group at the end of each of the fatty acid molecules reacts with an -OH (hydroxyl) group on the glycerol molecule.

Unlike single bonds, which are very flexible, double bonds are rigid. Fatty acids without double bonds are referred to as 'saturated' (because every carbon atom has a bond with a hydrogen atom). Fatty acids with one double bond are called 'mono-unsaturated' and those with many double bonds are called 'polyunsaturated'. The chains of unsaturated fatty acids are bent, so they take up more space. This looser arrangement means that unsaturated fats are less solid than the saturated ones, with their closely packed chains. This

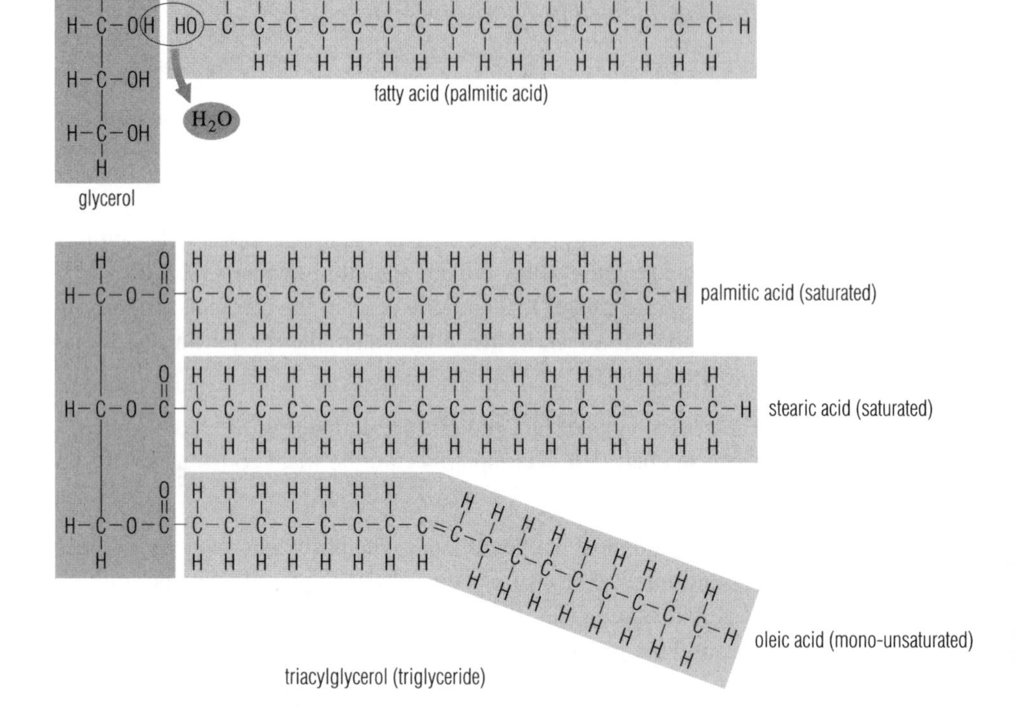

Figure 21 the formation of a triacylglycerol molecule by the successive addition of three molecules of fatty acid to a molecule of glycerol. Notice that a molecule of water is eliminated for each bond formed

means that fats, which are fairly solid, contain a higher proportion of saturated fatty acids than the relatively unsaturated, and therefore more liquid, oils. When sunflower oil is manufactured into sunflower margarine, hydrogen is added to the unsaturated oil (to take up the double bonds) which causes it to become more saturated and, therefore, more solid.

Because each triacylglycerol molecule is formed from glycerol and three carboxylic (fatty) acids there are many different combinations of fatty acids which are found, each producing fats and oils with different properties. Table 2 shows the chemical formulae of some of the common carboxylic acids (fatty acids) which are used in nature to make triacylglycerols.

Question 10

Lard is a fat which used to be commonly used to fry foods. Would lard be mostly composed of saturated, monounsaturated or polyunsaturated fatty acids?

Table 2

Name	Structure/formula	Type
Butyric	$CH_3(CH_2)_2COOH$	saturated
Capric	$CH_3(CH_2)_8COOH$	saturated
Lauric	$CH_3(CH_2)_{10}COOH$	saturated
Myristic	$CH_3(CH_2)_{12}COOH$	saturated
Palmitic	$CH_3(CH_2)_{14}COOH$	saturated
Stearic	$CH_3(CH_2)_{16}COOH$	saturated
Oleic	$CH_3(CH_2)_7CH{=}CH(CH_2)_7COOH$	monounsaturated
Linoleic	$CH_3(CH_2)_4CH{=}CHCH_2CH{=}CH(CH_2)_7COOH$	polyunsaturated
Linolenic	$CH_3CH_2CH{=}CHCH_2CH{=}CHCH_2CH{=}CH(CH_2)_7COOH$	polyunsaturated

Chocolate

Chocolate is made from cocoa, cocoa butter and sugar. Cocoa butter is made up of only a few different types of triacylglycerol molecule so can have a fairly sharp melting temperature. In fact, it can solidify in six different forms. This means the molecules can pack together in different ways leading to different types of crystal. One of these forms melts at 33.8°C. Thus it melts in your mouth (36.9°C) not in your hand. This form is also smooth and glossy. To persuade it to solidify in this form the chocolate is cooled and maintained at just below 33.8°C. It is also stirred so the fat crystallizes into very small crystals, which gives chocolate its velvety texture.

If chocolate is subjected to fluctuations in temperature, as in a shop window, a bloom develops on the chocolate. This is not mould but is the fat crystallizing out in different crystalline forms. Since cocoa butter is such a difficult fat to crystallize in the desired form, chocolate substitutes have to be used in cooking.

Vitamins and minerals

Vitamins and minerals are not structured in the same way as the macronutrients described above and it is not necessary for you to understand the chemical composition of them.

Creating foods with useful structural properties

Now that you have some understanding of the structure of the different nutrients within foods and some understanding of the chemical reactions, you can begin to see how these properties can be used in food product development.

With fibrous proteins, such as those found in meat, it is important that any food product requires a cooking process which will introduce water into the structure so that the meat will be softened for eating. Alternatively, another chemical may be used to soften the meat fibres, for example papain which is used in meat tenderizer products for barbecues. The box on 'Meat tenderizing' discusses this in more detail.

Globular proteins have a more complex structure, which gives them different properties. The most important property they have is that of 'denaturation'; on cooking the heat breaks the bonds between the protein chains, they reform in a different shape and cannot then be changed back. This property is exemplified when eggs are used in quiches and flans; they go into the flan as a liquid, are heated and set, thus holding together the other ingredients in the pastry case. Similarly with meringues, when the egg whites are beaten the protein albumin surrounds each air bubble and stabilises the foam; this is stabilised further by the addition of sugar. If heated, the protein coagulates and a solid foam, or meringue is made. A meringue cannot be made without sugar as it would not be sufficiently stable and would collapse. If overbeaten, however, too much air is incorporated and too much protein is denatured, making the protein films too thick and elastic. If this happens, when baked the air inside the air bubble would expand and break the cell wall before the protein had become coagulated by the heat of the oven, causing the meringue to collapse!

Gelatine is a form of protein, made from the collagen of skin, ligaments or bone which is hydrolized with alkali or dilute acid to give a solution which, on cooling, sets to a gel. This is important in cooking because it gives rigidity, for example when used as a covering for fruits in a flan case. Gelatine molecules are large, with a thread-like shape, and they are hydrophilic (water-loving). The structure is a double-helix, held with hydrogen bonding, which gives a three-dimensional structure which can hold liquid. When gelatine is added to a mixture, its long thin molecules are dispersed through the mixture, water molecules are attracted to the large surface area of gelatine; this causes the water molecules to lose their freedom within the mixture and so a more rigid mixture is obtained. Unlike denaturation, this process can be reversed. As a gelatine mixture is heated the hydrogen bonds are broken and the gel structure weakens; as it is cooled the bonds reform and the gel becomes more rigid again. The rigidity obtained will depend on the concentration of gelatine to liquid, as the concentration increases the rigidity will increase.

In sugar forms of carbohydrate, their main property is providing sweetness. Starch carbohydrate, on the other hand, has a useful property due to its reaction with liquid. Starch granules are insoluble in cold water, forming an unstable suspension. However, when starch granules are heated, in the presence of liquid, at certain temperatures (these vary according to the particular type of starch) the starch granules burst open releasing their starch into the liquid, causing it to thicken. This process is known as 'gelatinisation'. The changes are gradual and occur as shown in Table 3.

Table 3 the stages in gelatinisation

Stage 1	in cold water, the granules imbibe about 25-30% water, there is no change in the viscosity of the mixture
Stage 2	at about 65°C (this varies with different starches), the granules swell rapidly and take up a large amount of water. The typical granular structure disappears and the granules become swollen sacs, touching each other. At this stage, the shorter and more soluble starch molecules are leached out of the granules into the liquid; this change is irreversible.
Stage 3	as the temperature increases, more swelling takes place, and the granular structure breaks down. More starch is leached out and the viscosity of the liquid increases.

This is shown in the use of cornflour to make sauces. If the liquid is not stirred at stages 2 and 3, the starch released will coagulate and give a lumpy sauce! It is important, therefore, to stir the mixture constantly to ensure that the starch released is distributed evenly throughout the liquid and gives an evenly thickened mixture.

In food product development, the main property of fats and oils is their ability to form emulsions, a liquid-in-liquid mixture usually oil-in-water. If emulsions are left they will settle out into two layers, so in order to form a stable emulsion a third agent is added, an emulsifier or emulsifying agent, for example lecithin (a natural emulsifier found in egg yolk and some vegetable oils) or glycerol monosterate. Mayonnaise is an example of an emulsion, where oil droplets in water are stabilised by the lecithoprotein content of the egg yolk. Stabilisation of the emulsion can also be helped by 'homogenisation', in which the size of the dispersed fat globules is greatly reduced to a more or less uniform diameter.

Question 11

The process for making the lemon filling for meringue is:

- *Blend cornflour with water and heat*
- *Stir butter into the mixture*
- *Add lemon juice, lemon rind, sugar and egg yolks*
- *Stir the mixture then pour into the pie case*
- *When the topping is added, bake in the oven*

Explain the function of each of the ingredients used in the recipe.

Digestion of food

Introduction

To help you understand how the nutrients within food are used by the body, it is helpful for you to have some understanding of the human digestive system. The foods we eat are mainly composed of a complex mix of molecules, as you have seen. These are insoluble and too large to be taken up by cells in the body, so must be broken down into smaller, soluble sub-units, which can be taken up from the gut into the circulation and then delivered to the different parts of the body. These processes involve the action and interactions of several different organs which are collectively known as the digestive system.

The digestive system

The anatomy of the digestive system is shown in figure 1. The gut (also known as the digestive tract or gastrointestinal tract) is very long, about 4.5 metres in an average adult. Associated with the gut are a number of other organs that play an essential role in digestive processes, for example by producing digestive secretions. These organs, such as the liver and pancreas, are also shown in figure 1.

A number of distinct processes are involved in the assimilation of foods. The processes by which the components of food are broken down into simpler forms are collectively known as 'digestion', while the uptake into the body of the products of digestion and of small molecules and ions such as water, mineral salts and vitamins is known as 'absorption'. The elimination of faeces (waste matter together with unabsorbed materials) is known as 'excretion'. There is a final important process which goes on in the gut and that is one of defence. Along with the foods we eat, we ingest a vast number of potentially pathogenic microbes. The immune system of the gut has some unique features which help defend us against harmful microbes in our food.

Foods enter the gut through the mouth, which is involved in the initial processing of food. Chewing and lubrication with saliva aids swallowing. When food is swallowed it is carried down the oesophagus by gentle muscle action and gravity, until it reaches the stomach. This is essentially a reservoir where food is prepared for the main stage of digestion in the small intestine.

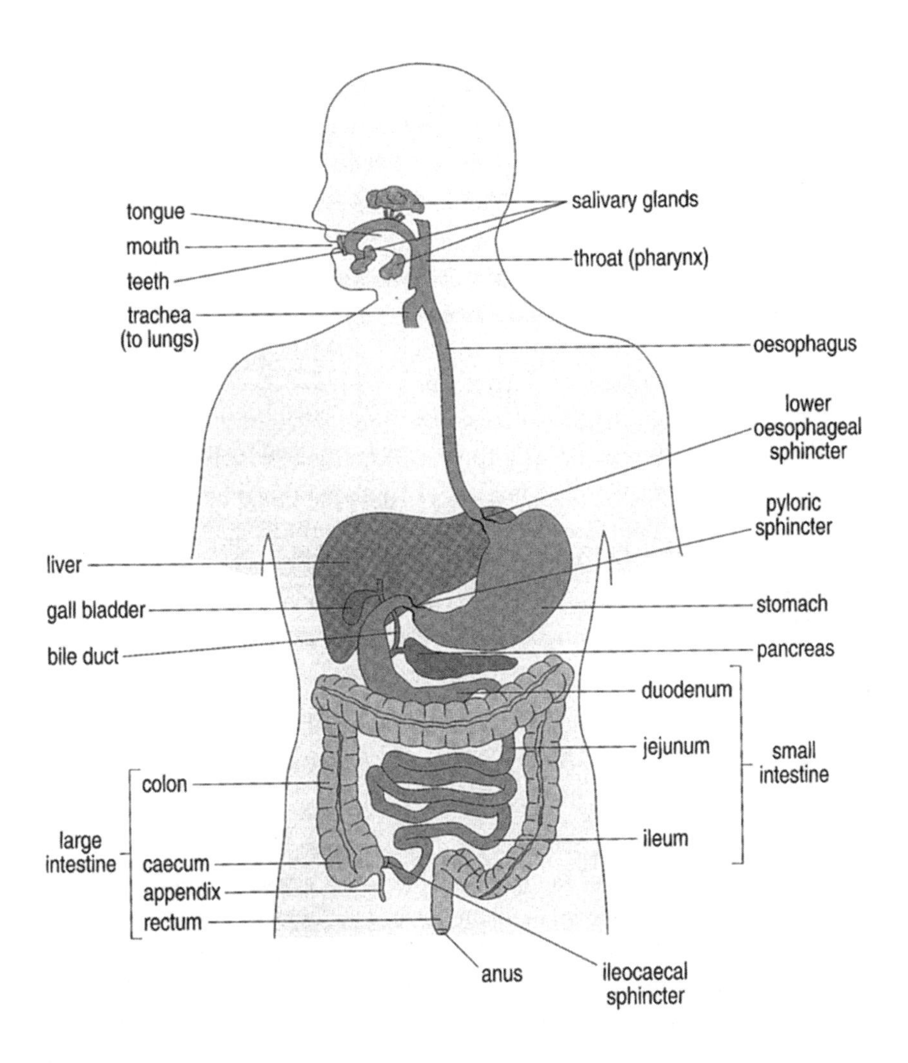

Figure 1 Anatomy of the digestive tract (gut) and associated organs

In the stomach the food is mixed with gastric juices, which are produced in the lining of the stomach. A flow of gastric juices is stimulated psychologically (for example when one thinks about food) and chemically (in response to food being tasted). However, the production of digestive juices may be inhibited by factors such as excitement, depression, anxiety and fear. The mixing of food and gastric juices takes place due to vigorous muscular activity in the lower region of the stomach about 20 minutes after eating a meal. The gastric juice contains hydrochloric acid and so is acidic, with a pH of about 1.5. This dilute acid aids the breakdown of the chemicals in the food and also kills off any bacteria. The gastric juice also contains two enzymes: pepsin, which starts breaking down the protein component of the food into individual amino acids, and rennin, which aids the coagulation of any milk in the food. The stomach has an outlet valve, which opens at intervals, so allowing the food and gastric juices to pass into the first part of the intestines, the duodenum.

The main stage of digestion occurs on passage through the small intestine using a new set of compounds provided by:

Bile	a range of chemicals produced in the liver and stored in the gall bladder
Pancreatic juice	a group of chemicals secreted by the pancreas
Intestinal juice	chemicals secreted by the lining of the intestine

These three juices are basic and so neutralize the acidity of the gastric juice. They also provide new enzymes, such as peptidases, amylases and maltases, and lipases.

Question I

From the names of these enzymes, and what you know of the structure of nutrients, can you suggest which nutrients each compound will break down?

The bile does not contain any enzymes, only bile salts. These are emulsifying agents that stabilize the formation of emulsions from the fats and fatty acids. In this more dispersed state the enzymes known as lipases can get at the fat molecules more easily and break them down. The digestive process is almost complete after the food has been in the small intestine for four hours. All the protein has been broken down into amino acids, all carbohydrates, except dietary fibre, will have been broken down into simple sugars and fatty acids will have been produced from the fats. During this stage these basic building blocks are absorbed into the body. The walls of the long small intestine are folded into finger-like projections, which contain both blood capillaries and lymph vessels. Water-soluble materials, such as amino acids, sugars, minerals and some vitamins, dissolve into the blood in the capillaries and are then transported away through the bloodstream to other areas of the body. Fatty acids, which do not dissolve in the blood, are absorbed into the lymph vessels, where they are converted back into triacyglycerols. These new triacylglycerols contain combinations of fatty acids that are more suitable for use in the body. They are then emulsified so that they can be carried round in the blood stream. Other water-insoluble materials, such as cholesterol and some vitamins, are also emulsified with the fats, for transportation.

Table 1 summary of digestive activities

Gut region or digestive organ	Digestive secretions	Functions
mouth (jaw, teeth and tongue)		chewing; initiation of swallowing reflex
salivary glands	salts and water	moistens food
	mucus	lubricates food
	amylase	begins beakdown of polysaccharides
oesophagus		moves food to stomach by peristals
stomach		stores, mixes, dissolves and begins main digestion of food; regulates passage of partially digested food (chyme) into small intestine
	hydrochloric acid (HCl)	dissolves food; activates pepsinogen; kills microbes
	mucus	lubricates and protects epithelial surface
	pepsin (from pepsinogen)	begins protein digestion
pancreas	bicarbonate (HCO_3^-)	neutralizes HCl entering small intestine from stomach
	proteolytic enzymes, amylase, lipase and nucleases	digest proteins, polysaccharides, lipids and nucleic acids, respectively
liver		produces bile (contains bile salts, HCO_3^-, waste products)
	bile salts	emulsify lipids
	HCO_3^-	neutralizes HCl entering small intestine from stomach
gall bladder		stores bile between meals and releases it into small intestine as required
small intestine		mixing and propulsion of contents by peristalsis
	disaccharide-splitting and proteolytic enzymes which are not secreted but present in brush borders of epithelial cells	complete digestion of carbonhydrates and protein
	enzyme (enterokinase) that activates a pancreatic proteolytic enzyme (trypsin)	
		absorption of products of digestion
colon		storage and digestion of non-digestible matter; mixing and propulsion of contents by peristalsis
	mucus	lubrication of faeces
rectum		defaecation

Approximately seven to nine hours after the food was eaten, any that has not been digested and absorbed in the small intestine passes through another valve into the wider, but shorter, large intestine. No new enzymes are produced, but the large intestine is rich in bacteria which break down some of the remains of the food. In this way small but important amounts of vitamins, such as vitamin K, are produced and absorbed by the body. However, the main function of the large intestine is to recover water from the fluid mass, so that after about twenty hours, by the time it reaches the anus, it is in a semi-solid form. Each day between 100-200g of moist faeces may be produced, consisting of undigested food material, such as dietary fibre (non-starch polysaccharide), residues from digestive juices, living and dead bacteria from the large intestine and also water.

Table 1 summarises the digestive processes described above.

Note: some people are unable to digest lactose, a sugar found particulary in milk. This is known as 'lactose intolerance' and requires the avoidance of milk and milk products.

Food and nutrition

Introduction

Around the world communities of people eat foods different from our own yet, with sufficient to eat, we all grow and are healthy. The reason for this is that the wide range of foods available all contain the same basic components – nutrients – which supply the body with what it needs for growth, energy, repair and maintenance, and substances which regulate bodily processes. The major classes of nutrients are proteins, carbohydrates, fats, vitamins and minerals. It is interesting to note that all foods, irrespective of whether they have an animal or vegetable source contain mainly these nutrients. To stay alive we need to take in these five nutrients, together with water and oxygen.

Table 1 shows the approximate percentage of body weight of each of the nutrient groups in the average person's body.

Table 1 percentage of body weight of the main nutrients

Nutrient	% of body weight
Proteins	15-20
Fats	3-20
Carbohydrates	1-15
Small organic molecules	0-1
Inorganic compunds	1
Total solid material	20-40
Water	60-80

Figure 1 shows the functions in the body of each of these nutrients, notice how several different nutrients perform the same function.

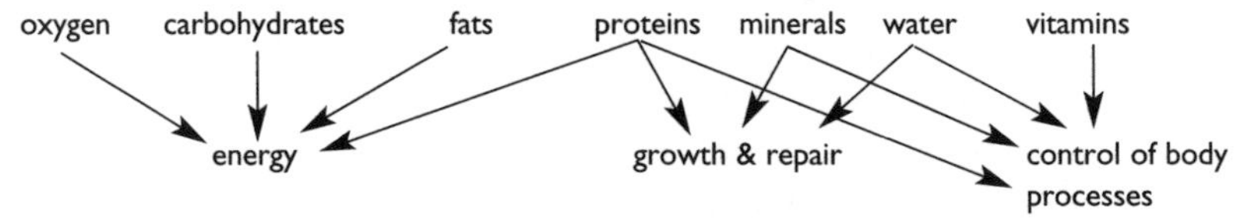

Figure 1: shows the relationship between the five nutrients, oxygen and water, together with their major functions.

The 'macronutrients', so called because the body needs relatively larger quantities, are protein, carbohydrates and fats. 'Micronutrients', needed in smaller amounts, are vitamins and minerals. This section will consider foods classified according to their nutritional properties.

Protein

All proteins are made up of carbon, hydrogen, oxygen and nitrogen; most also contain sulphur and some contain phosphorus. The section on 'The structure of nutrients' discusses this further.

The body requires protein mainly for growth and repair, but if there is a lack of energy in the diet protein can be converted into carbohydrate and used for energy. In children the actions of protein can easily be seen, as their bodies, muscles and bones grow and develop. However, adult bodies are also constantly being renewed (for example old skin sloughs off and new skin replaces it, hair and nails continue to grow), so the need for protein continues.

Protein intakes in Britain are generally adequate for our bodily needs. Extra dietary protein is needed by people who are suffering from injury, infection, burns and cancer, as all these can result in an increased loss of protein.

Question 1

At what stages of life might there be a requirement for extra dietary protein? Think about the function of protein in the body and when this might be particularly needed.

A lack of sufficient protein in the diet leads to a slowing down, or stopping, of growth. It then causes a loss of tissue renewal, which in internal organs can have drastic consequences, for example causing failure of the digestive system which, in turn, causes diarrhoea and loss of water. Muscle wasting and anaemia follow. In the developing world diets are often deficient in protein, as well as having an overall inadequate energy content. Protein-energy malnutrition (or protein-energy deficiency) is a condition where the diets of adults and children are lacking in a range of nutrients and overall food intake is too low for their bodily requirements. Marasmus is a severe form of protein-energy malnutrition and is a condition often seen in famines. Although low-level marasmus can occur throughout a vulnerable population, only the most severe cases are noticed. Once the body's fat stores are depleted, muscle and organ protein is broken down to provide energy. Protein-deficiency in children can also cause a condition called kwashiorkor. This can occur when children are weaned from breast milk onto protein-poor foods such as cassava (a root vegetable) or green bananas.

Protein can be found in both animal and plant foods. The animal sources are:

- meat
- fish
- cheese

- eggs
- milk.

The proteins in these foods are similar in structure and composition to those of the human body, so can be more easily assimilated.

Plant sources include:

- cereals
- nuts
- lentils
- peas
- beans.

The composition of these, however, means that they do not supply the human body with all its protein needs (essential amino acids) and they must be eaten in combination, for example beans on toast, to be of value. This is because the food value missing in one vegetable protein can be found in another, and when eaten together they combine to form a complete protein which the body can then use. This is discussed further in the section 'The structure of nutrients'.

The soya bean is an exception to this; although a plant source of protein it provides protein which is directly of value to humans. Many products have now been developed from soya, also known as textured vegetable protein (TVP), to resemble meat and to provide high-value protein to vegetarians.

Table 2 shows the protein content of some common foods.

Table 2 the protein content of some common foods

Animal sources	**protein content %**	**Plant sources**	**protein content %**
Cheese (cheddar)	26	soya flour (low-fat)	45
Bacon (lean)	20	soya flour (full fat)	37
Beef (lean)	20	peanuts	24
Cod	17	bread (wholemeal)	9
Herring	17	bread (white)	8
Eggs	12	rice	7
Beef (fat)	8	peas (fresh)	6
Milk	3	potatoes (old)	2
Cheese (cream)	3	bananas	1
Butter	less than 1	apples	less than 1

Other foods also contain protein, but in amounts which make their contribution to the diet insignificant.

> ***Activity 1***
> Think about your own food intake, approximately how much of it is composed of high-protein foods. Is this appropriate for your stage in life? Do you think that you should be consuming more or less protein?

Carbohydrate

There are a number of different forms of carbohydrates but all are composed of carbon, hydrogen and oxygen, and all come from plant sources. The chemical structures of the different forms are discussed in the section 'The structure of nutrients. There are two types of carbohydrates – starch and sugars. Starch is the main energy storage material for plants and can be found in potatoes and in seeds such as cereal grains, peas and beans. Starch forms the main dietary carbohydrate for much of the world's population. Sugars are made by plants as sources of energy and stored around their seeds, in the flesh of fruits. They can also be found in honey.

Sugars can be further classified into two groups:

- intrinsic sugar – defined as sugar that is contained within the structure of the foodstuff (i.e. in the tough walls of the plant cells which make up the food material)
- extrinsic sugar – defined as any sugars not incorporated into the structure of the food, such as sugars extracted from fruits and those in honey and table sugar

Carbohydrates, in the form of starch and sugars, provide the body with energy. In fact, they are the main energy source in most diets, particularly those in developing countries where they can provide 80-90% of the energy requirement. If more carbohydrate is eaten than the body requires for energy, it converts the excess to fat and stores it. If this continues it can cause obesity, which in turn can lead to heart disease. In addition, an excess of sugar can cause dental caries and may lead to mature-onset diabetes.

> ***Activity 2***
> Think about how much of your diet is made up of carbohydrate foods. Is this mostly starch, or sugar?

There is a further group of carbohydrates, known as NSP (non-starch polysaccharides), or dietary fibre, which consists of cellulose and pectin. These are forms of starch carbohydrates but they are not classed as nutrients because the human body is unable to break them down in digestion, so cannot absorb and utilise them. NSP does have a value in the diet, however, as it assists in the excretion of waste from the body. Eating insufficient dietary fibre can lead to bowel diseases.

Question 2

If non-starch polysaccharide cannot be digested and assimilated, why is it important in the human diet?

Activity 3

From table 3, identify in your own diet the sources of non-starch polysaccharides. Do you think that you have a sufficiently high intake? If you need to increase your intake, how could you do this?

Sources of carbohydrate are varied, but consist mainly of:

Table 3 Common sources of carbohydrates

Starch	Sugars	Non-starch polysaccharides
Cereals -wheat, maize, rye, pasta, oats	Jam	Wholemeal products – flour, bread, brown rice
Cereal products – flour, bread, biscuits	Honey	Vegetables
Rice	Dried fruits	Oats
Pasta	Fruit	
Cassava	Chocolate	
Yam	Sugar	

Question 3

The two main types of digestible carbohydrates are starch and sugar. What are the different types of digestible starch and the different types of sugar?

Fats and oils

Generally fats are solid at room temperature, whilst oils are liquid. Another term for fats which you may come across is 'lipids'. Fats and oils, like carbohydrates, are composed of carbon, hydrogen and oxygen but the proportion of oxygen is lower than in carbohydrates. The structure of fats and oils is explained in detail in the section 'The structure of nutrients'.

Fats provide the body with high levels of energy; they contain twice as much as protein or carbohydrates. The body stores excess energy as a layer of fat under the skin, which helps to insulate the body and protect internal organs. Although an excess of fat in the diet leads to increased body fat, weight gain and associated health problems, fats are an essential component of the human diet as they contain 'essential fatty acids' (discussed in 'The structure of nutrients') and certain vitamins (A, D, E and K) that are required by the body and cannot be obtained from other sources.

It is difficult to ascertain the effects of a fat-deficient diet, as most diets deficient in fat are also deficient in protein and total food content, although one obvious result is a lack of the fat-soluble vitamins. A greater problem is an over-consumption of fats, resulting in weight gain and related health problems. In addition, if high levels of saturated fats (discussed in 'The structure of nutrients') are eaten these could lead to coronary heart disease due to high levels of cholesterol.

The sources of fat in the diet can be described in several ways. A distinction can be made between 'visible fats' and 'invisible fats'. The first refers to those products we can see as fats, such as butter, margarine, lard and oils; the second to those we cannot see, such as the fat in meat, cheese, cakes and chocolate.

Activity 4

It is easy to identify 'visible fats', but think about your own diet and try to identify the 'invisible fats'. Think about the ingredients of the foods that you eat, do they contain butter, margarine, oil? Think about how the foods are cooked, are they fried or roasted? Think, also, about the manufactured foods that you eat and check the labels to find out if they contain hidden fats.

Fats can also be divided into 'animal' and 'vegetable' sources. A similar result would be obtained from dividing them into 'saturated fats' and 'unsaturated fats'. Although not all animal fats are saturated and not all vegetable ones are unsaturated the majority in each case are. Table 4 shows some of the major sources of animal and vegetable fats.

Table 4 the major sources of animal and vegetable fats

Animal fats (mainly saturated)	**Vegetable fats** (mainly unsaturated)
Meat products – beef, pork, lamb	Vegetable oils – olive, sunflower, corn
Dairy products – butter, cheese, milk, eggs	Vegetable margarines
Fatty fish – salmon, mackerel, tuna, herrings – these contain unsaturated fats	Nuts
Suet	
Ghee	

Table 5, below, shows the fat content of a range of foods.

Table 5 the fat content of a range of foods (listed in descending order of fat content)

Food	**fat content %**	**Food**	**fat content %**
Lard	99	beef (rump steak)	14
Margarine	81	eggs	10.9
Butter	82	milk	3.8
Cream cheese	47	bread (white)	1.7
Cheddar cheese	34	rice	1
Pork sausage	32	haddock	0.6
Herring	19	potatoes	0

Activity 5

In your diet, is the main source of fats animal or vegetable? How healthy do you think this is?

Question 4

Why is too much fat in the diet harmful?

It is worth mentioning cholesterol here. Cholesterol is a type of lipid that is present in body tissues and is found in some foods. It has been perceived to be a nutrient to be avoided because of its links with heart disease. However, cholesterol is important in the body as a component of cell membranes; it assists in the manufacture of vitamin D, some hormones and bile salts. The body actually manufactures its own cholesterol and only 20-25% comes from dietary sources. Under normal circumstances, if dietary intake is high there is a corresponding reduction in the body's own synthesis. However, the total intake of fat and the types of fat eaten can have an effect on blood levels of cholesterol.

Foods with high levels of cholesterol include eggs, liver, kidney and foods containing saturated fats.

Vitamins

Vitamins are a diverse group of organic compounds known as micronutrients, as they are only required by the body in small amounts. However, they are essential for the maintenance of body functions and good health.

Vitamins were discovered and named before their detailed chemical structures were known and so they tend to be referred to by a letter as well as by a chemical name, for example vitamin C is ascorbic acid and vitamin D is cholecalciferol. The use of letters ran into problems when it was discovered that the original vitamin 'B' was, in fact, a collection of quite distinct vitamins, and so each was assigned a number, B_1, B_2 etc. This system was finally thrown into chaos and abandoned when it was discovered that a number of vitamins were identical, thus we no longer have vitamins F to J, or B_3 to B_5.

There are two major groups of vitamins. Those known as 'fat-soluble' (vitamins A, D, E and K), as they dissolve in fat and are found in fatty foods. The body can build up stores of fat-soluble vitamins in the liver and it is possible to have levels which are too high in the body, and therefore toxic. Excess intakes of vitamin A can cause liver and bone damage, hair loss, double vision, vomiting and headaches. Vitamin overdose in pregnant women can cause birth defects in their unborn children. The other group of vitamins are water soluble, they are vitamins B (this is a group of vitamins) and C. The body takes what it needs from the foods ingested and any excess is removed from the body via the urine. This means that a regular intake is necessary.

Each vitamin so far identified plays a specific role in maintaining the body in good health. The absence of a vitamin from the diet, or its insufficiency, leads to specific symptoms.

A major role played by some vitamins is that of an antioxidant. Vitamins C, E and A (and carotene, which can be converted in the body to vitamin A) are antioxidants. They play an important protective role by limiting the action of harmful substances produced by some of the chemical reactions that take place in the body. These harmful substances are known as 'free radicals', which are atoms (or groups of atoms) which contain unpaired electrons and so are unstable and highly reactive. They acquire electrons from other molecules that are around them, thus creating another unpaired electron in that molecule and so on, causing a chain reaction. This chain reaction can cause considerable damage to living material. Antioxidants 'mop up' (i.e. through supplying electrons) free radicals before they can cause too much damage. Free radicals have been implicated in many human diseases and disorders although their exact role in many of these has still to be fully understood. Many pollutants generate free radicals, as does smoking, and it is this that is thought to be their link with diseases such as cancer.

The major vitamins, together with their function, deficiency symptoms and sources in the diet, are shown in Table 6. Vitamins can be found in a wide range of foods, Table 6 shows only the foods containing amounts that can make a significant contribution to the diet.

Question 5

Which vitamins are most commonly added to foods? (You can find this out from looking at a range of processed foods). Why do you think that these particular vitamins are added?

Question 6

Which vitamins are dangerous if high dosages are taken?

Vitamin deficiencies in the developed world are now rare. However, in some cases although there may be enough of a particular vitamin in the diet, inadequate amounts of it may actually be absorbed. In addition, there are bacteria in the colon which produce several vitamins. Antibiotic treatment can destroy these bacteria, so this would reduce the amounts of vitamins that are absorbed. A diet high in dietary fibre (NSP) can also prevent the absorption of some vitamins.

Minerals

Minerals are inorganic compounds, again required in small amounts in the body but essential for the maintenance of bodily functions and good health. Minerals can be further divided into the 'major minerals', those required in *relatively* larger quantities (calcium, phosphorus, sodium, iron), and 'trace elements' required only in smaller amounts (such as zinc, copper, selenium).

Minerals have many different roles in the body; these can be summarised as follows:

Table 6 vitamins needed in the human body

Name of vitamin	Main sources	Functions in the body and effect of deficiency
Fat-soluble vitamins		
A, or retinol	milk, dairy products, margarine, fish-liver oil. Can also be made from carotenes found in green vegetables and carrots	necessary for healthy skin, normal growth and development; antioxidant. Deficiency will slow down growth and may lead to disorders of the skin, lowered resistance to infection, disturbances of vision (such as night blindness) Can be toxic if too much is consumed
D, or cholecalciferol	margarine, buttermilk, fish-liver oils, oily fish	necessary for the formation of strong bones and teeth. Deficiency may cause bone diseases (such as rickets) or dental decay.
E, or tocopherols	plant seed oils	antioxidant
K, or naphtho-quinonesgreen	vegetables	assists blood clotting
Water soluble vitamins:		
B group of vitamins:		
B_1, or thiamine B_2 or riboflavin Niacin B_6 or pyridoxine Pantothenic acid Biotin	bread and flour, meat, milk, potatoes, yeast extract, fortified breakfast cereals	involved in many of the reactions that release energy from food. Deficiency causes loss of appetite, slows growth and development and impairs general health, severe deficiency leads to diseases such as pellagra or beriberi.
B_{12} or cobalamin	offal, meat, milk, fortified breakfast cereals	Necessary for the formation of nucleic acids and red blood cells. Deficiency may lead to certain types of anaemia
Folic acid	potatoes, offal, green vegetables, bread, Marmite, fortified breakfast cereals	has a preventative role in the occurrence of neural tube defects
C, or ascorbic acid	green vegetables, citrus fruits, potatoes, blackcurrant and rosehip syrup	necessary for the proper formation of teeth, bones and blood vessels; antioxidant. Deficiency retards the growth of children and, if prolonged, may lead to scurvy

- They form essential structural components of cells and tissues; for example, sodium, potassium, calcium and chlorine, in their ionic forms, are found in all cells and extracellular fluids, while calcium, phosphorus and magnesium salts are major components of bones and teeth
- The ions of several minerals (sodium, potassium, calcium) play an essential role in intercellular communication and in the transport of small molecules across cell membranes by active transport. Calcium ions also act as intracellular messengers which, among many other things, trigger neurotransmitter release at nerve terminals
- Minerals are also essential components of many important molecules including some hormones (for example iodine is an essential constituent of the thryroid hormones); haemoglobin (in which iron is bound to haem) and many enzymes.

Table 7 shows some of the major minerals and trace elements, their main food sources and their functions in the body.

Question 7

Which processed foods have minerals added to them? Why do you think this is?

The fact that minerals are needed only in small amounts, and because they are abundant in foods, people who eat a varied diet are unlikely to develop mineral deficiency.

Dietary fibre and water

Dietary fibre, or non-starch polysaccharides, was referred to earlier, in the section on carbohydrates. It is mentioned again, together with water, as a reminder that these two are not classed as nutrients, because they are not absorbed into the body, although they are essential for good health.

Question 8

Why is dietary fibre, or non-starch polysaccharide, essential for good health?

Nutrients in foods

This section has looked at individual nutrients, but most foods are composed of a mixture of nutrients. Think of a typical traditional English breakfast – orange juice; bacon, egg and tomatoes; toast, butter and marmalade; coffee.

The orange juice is mainly water (88%) with about 0.5% protein and 10% carbohydrate. It is a particularly rich source of vitamin C and contains small amounts of minerals and other vitamins.

Table 7 major mineral elements and trace elements needed by the body

Mineral	**Main food sources**	**Functions in the body**
Calcium	milk, cheese, bread and flour (if fortified), cereals, green vegetables	present in bones and teeth, necessary for blood clotting, muscle contraction and nerve activity
Phosphorus	milk, cheese, bread and cereals, meat and meat products	present in bones and teeth, essential for energy storage and transfer, cell structure, cell division and reproduction
Sulphur	protein-rich foods, such as meat, fish, eggs, milk, bread and cereals	present in the body in proteins, important for cross-linking proteins
Sodium	main source is salt (sodium chloride), used in food processing, cooking and at the table; bread and cereal products are the main sources in processed foods	present in body fluids as sodium ions, essential for maintenance of fluid balance in the body and for nerve activity and muscle contraction
Chlorine	see above (sodium chloride)	present in gastric juice and body fluids as chlorine ions
Potassium	widely found in vegetables, meat, milk, fruit and fruit juices	present in cell fluids as potassium ions, similar role to sodium, but they are not interchangeable
Iron	meat and offal, bread and flour, potatoes and vegetables	essential component of haemoglobin of blood cells
Magnesium	milk, bread and other cereal products, potatoes and vegetables	present in bone and cell fluids, needed for activity of some enzymes
Zinc	meat and meat products milk and cheese, bread, flour and cereal products	essential for the activity of several enzymes involved in energy changes and protein formation
Trace elements		
Cobalt	liver and other meat	required for formation of red blood cells
Copper	green vegetables, fish and liver	component of many enzymes; necessary for haemoglobin formation
Chromium in	liver, cereals, beer, yeast	contained in all tissues; may be involved glucose metabolism
Fluorine	tea, seafood, water	required for bone and tooth formation
Iodine	milk, sea food, iodized salt	component of thyroid hormones
Manganese	tea, cereals, pulses, nuts	forms part of some enzyme systems
Molybdenum	kidney, cereals, vegetables	enzyme activation
Selenium	cereals, meat, fish	present in some enzymes; associated with vitamin E activity

The bacon contains mainly fat and protein. The fat is quite obvious as a layer between the skin and muscle (lean meat). The muscle contains about three-quarters water and one-quarter protein, together with small amounts of fat, minerals and vitamins. Both fibrous and globular proteins are present, with the former predominating. The main minerals are sodium, potassium, calcium and iron, although meat is also a good source of zinc. The key vitamins in bacon are niacin, thiamine and riboflavin.

The egg is an interesting food because it is designed to contain all the nutrients that a developing chick embryo will need. The outer shell is made of calcium carbonate, but is porous enough to allow sufficient oxygen to reach the chick. The egg white accounts for about 60% of the mass of the egg and is one-eighth protein and seven-eighths water. The main protein, ovalbumin, is of the fibrous type. It provides a store of amino acids for the growing chick. Globular proteins are also present in smaller amounts. Dissolved in the water are small amounts of salts and the vitamin riboflavin. In the centre of the egg is the yolk; it is one-third fat, half water and one-sixth protein, again with salts and vitamins. Associated with the protein are relatively large amounts of phosphorus. There is virtually no carbohydrate in eggs because the chick gets all its energy from fats.

Tomatoes contain mainly water, 95% by mass, with about 3% carbohydrate, 1% protein and no fat. The main minerals are calcium, sodium and iron. The main vitamins are vitamin C and niacin, together with small amounts of riboflavine and thiamine; however, tomatoes are a reasonable source of vitamin A.

Bread comes from flour, which in turn comes from wheat. If the flour used to make the bread is 'wholewheat', 'wholemeal' or 'wholegrain' then it contains the outer layer of wheat, the bran, which is rich in vitamins and contains about half the minerals in the grain of wheat. The bran is mainly cellulose, which is indigestible by humans, and so provides dietary fibre. The wheatgerm is rich in fats, protein, vitamins (in particular vitamin E and thiamine) and iron. Most of the grain is taken up by the endosperm, which contains starch granules embedded in a matrix of protein. About 7-15% of the endosperm is protein, mainly fibrous. The wheat is milled and made into flour, which may involve removing various parts of the whole grain, such as the bran and the germ. If this happens the flour (and bread) loses colour (white flour, white bread), is more refined and some of the nutrients are lost. To counteract this, iron, thiamine and niacin are often added to the flour.

The butter or margarine spread on the toast is a visible fat, and the marmalade is made from oranges and sugar (about 67%). The pectin that helps the marmalade to set is a polysaccharide (starch).

There is little nutritional value to black coffee. The average cup of fairly strong coffee contains potassium, caffeine and niacin. Caffeine is not a nutrient in that it is not used by the body for growth or energy. If whole milk is added to the coffee then protein, fat and some carbohydrate are added. Calcium is the main mineral to be added, together with trace amounts of other vitamins and minerals. If skimmed milk is used the nutrients remain the same, except that a lower amount of fat will be in the milk. If sugar is also added to the coffee, then carbohydrate in the form of sugars is also added.

Question 9

What nutrients are provided by the typical English Sunday lunch – roast beef, Yorkshire puddings, potatoes, carrots, peas and gravy?

Energy in foods

Energy is required for sustaining all forms of life on earth. Energy in foods in measured in kilocalories (kcals) or the metric equivalent, kilojoules (kJ); one kilocalorie is approximately 4kJ. A kilocalorie is the amount of heat required to raise the temperature of one kilogram of water by one degree centigrade,

All foods contain energy, but in varying amounts according to their nutritional composition. One gram of protein and carbohydrate each contains approximately 4 kcals (16kJ), whilst one gram of fat contains 9 kcals (36kJ). Therefore, depending upon the amount of protein, carbohydrate and fat in each food, its energy (or calorie/Joule) value will vary.

During the digestion and absorption of food, energy is released in small amounts at various stages. The energy produced is used to:

- maintain internal and involuntary body processes, such as keeping the heart beating, the lungs breathing and the eyelids blinking
- maintain chemical reactions within the body
- maintain muscle tension
- maintain body temperature
- allow for voluntary physical actions, such as walking, moving and working.

After food has been eaten, the various processes it is subject to before it is utilized, such as digestion and absorption, result in the release of energy. Most of this appears as heat. This is known as **thermogenesis**, corresponding to about 10% of the energy content of food. Excess energy is converted to fat and stored in the body.

The term 'metabolism' covers all the chemical reactions that are going on in the body, and the metabolic rate of an individual who has not eaten for 12 hours, is at rest and at a comfortable temperature, is known as the **basal metabolic rate**. This is not a measure of the lowest metabolic rate, which occurs during sleep. Even when you are sleeping your body needs considerable quantities of energy to continue its essential processes, such as keeping the heart beating and the organs functioning. However, some muscles must constantly be ready to contract in response to stimuli from the nervous system, and thus energy is continuously needed to keep the muscles in a state of tension. The basal metabolic rate is affected by size, gender, age, rate of growth and even by climate, hormonal activity and the amount of sleep a person gets.

Table 8 lists, in kiloJoules, the average basal metabolic rates for different groups of people.

Table 8 some average values of basal metabolic rate

Group/Age	weight/kg	average rate/kj per day
Infant, 1 year	10	2,100
Child, 8 years	25	4,200
Woman, adult	55	5,400
Man, adult	65	6,700

The average basal metabolic rate for men and women accounts for about two-thirds of the total energy required by the body. The basal metabolic rate varies with age, falling off faster for men that for women as they get older. It also varies with climate, being reduced by 5-10% in very cold or very hot climates.

Whenever we do any kind of activity, be it standing up, walking, running, climbing, gardening, playing sport or going about our work, we use muscles that need energy. This energy is over and above that involved in the basal metabolic rate. Unfortunately, muscles are not very efficient machines and only 15-20% of the energy ends up doing the required work.

Table 9 average energy expenditure per minute for a range of activities

Activity		Average energy expenditure KJ per minute
Everyday activities:		
Sleep		5
Sitting		6
Standing		7
Washing, dressing		15
Walking slowly		13
Walking moderately quickly		21
Walking up and down stairs		38
Light effort:		
Most domestic work	}	
Golf	}	
Lorry driving	}	10-20
Carpentry	}	
Bricklaying	}	
Moderate effort:		
Gardening	}	
Tennis, dancing, jogging	}	
Cycling, up to 20km per hour	}	21-30
Digging, shovelling	}	
Agricultural work (manual)	}	
Strenuous effort:		
Coal mining, steel furnace	}	
Squash, cross-country running	}	over 30
Football, swimming (crawl)	}	

Table 9 shows some typical energy requirements for a range of activities in kiloJoules per minute of activity. These include an allowance for thermogenesis and basal metabolic rate.

Activity 6

From the above table calculate approximately your own energy use over an average 24-hour period.

Question 10

What are the factors that affect an individual's need for energy?

Clearly, if we take in more food than the body needs it will be stored and we will put on weight. About 45% of males and 36% of females in the UK are obese (above the upper level of weight considered desirable by the British College of Physicians). Alternatively, if not enough food is consumed, the body starts to convert its reserves of fat into water and carbon dioxide to provide energy, leading to weight loss. Both of these, weight gain and weight loss, have implications for health.

Maintaining an **energy** balance is important, so that there is neither unintended weight gain or weight loss. Even a small increase in kilocalorie intake, of 10 kcals a day, can produce a weight gain of 1lb (0.5 kg) over a year. Most people maintain a reasonably accurate energy intake-energy expenditure balance over a period of time without any effort; the body seems to regulate itself. However, if required, energy balance can be achieved either by amending food intake or by increasing or decreasing the amount of physical activity undertaken.

The amount of energy in individual foods can be found in approved food tables, such as McCance and Widdowson or the Manual of Nutrition.

Using information from the tables 8 and 9 above, it is possible to determine how much energy we need, on average, for our lifestyles and so can determine how much food we need to take in. Table 10 shows such a calculation for an average person's day.

This shows that the food intake for this person should provide about 11000 kJ per day to maintain energy balance. For most people, they often achieve this balance over a period of a few days, some days eating a little more or less, some days expending more or less energy.

Table 10 energy used in an average person's day

Activity	KJ
9.5 hours sitting	3420
2 hours standing	840
2 hours gentle walking	1560
0.5 hour going up and down stairs	1140
0.5 hour washing, dressing	450
1 hour light domestic activity	600
8 hours sleep	2400
0.5 hour squash	900
Total	11310

Activity 7

List all the foods you ate in one day, use yesterday as an example. Try to calculate the energy content of those foods. If they were manufactured foods this information should be on the label. If they were fresh foods you can calculate approximate energy values, or use food tables to check. How close was your energy intake to your energy expenditure?

Activity 8

Knowing the energy available per gram for the various kinds of nutrient, it is possible to calculate the energy value of a food, providing its composition is known. For example, the energy content of milk is calculated below.

Nutrient	amount per pint of milk	average available energy per gram of nutrient	energy available from the nutrient, in a pint
Protein	18.9g	17kJ	321kJ
fate	22.8g	37kJ	844kJ
carbohydrate	26.8g	17kJ	460kJ
			1625kJ

Using a similar tabular method, calculate the energy content per 100g of three foods from your own cupboard. Use the nutrition label to find out the amount of protein, fat and carbohydrate per 100g in the food.

The effects of preparation and processing on nutrients

The information given above all relates to food in its raw state, but most foods have to be prepared and processed before they can be eaten. Preparation may be simple, for example, the peeling of vegetables, or more complex, such as the primary processing of turning wheat into flour. In each case, however, some nutrients are destroyed or lost. Similarly, when applying heat to food this can lead to some loss of nutrients. Microwave cooking, because of its short cooking times and use of small amounts of water, is one cooking method which causes relatively little nutritional loss.

Freezing, providing the food is fresh when frozen and the temperature is kept below -18^{0}C, causes little nutritional loss. However, when the food is thawed and cooked, nutritional loss will be as for fresh foods. Commercially frozen foods generally keep high levels of nutrients, as the time from harvesting to freezing is short.

Protein, when heated, is denatured, that is it changes its chemical structure. This renders it less easy to digest and so less available for use by the body.

Carbohydrate is more stable and tends to be unaffected by preparation or processing.

Fats and oils themselves do not change when used in preparing and processing food, although their state may change, for example solid fats will melt to become liquids when heat is applied. However, when fats and oils are used in the processing of other foods they alter the nutritional properties of the other foods, by adding energy. Any food which is fried or roasted will have an increased energy content, so care must be taken if energy consumption is of concern.

Vitamins are most affected by preparation and processing, as they are reactive with oxygen, water and heat.

Vitamin A is mostly stable, although if high temperatures are used there may be some loss, for example when fats are used for frying. Foods containing vitamin A also need to be stored in cool, dark, airtight containers as prolonged storage with exposure to light and air could lead to loss.

Vitamin B-complex are all water-soluble and sensitive to heat. This means that if water is used to prepare foods, for example in soaking, or to process them, for example in boiling and stewing, there will be vitamin B loss. When heat is applied, particularly if conditions are alkaline, as when sodium bicarbonate is added to mixtures or to cooking water, then vitamin B loss can be great. Foods processed by canning or bottling will have lost B-vitamins, due to the high levels of heat used to kill off harmful bacteria.

Vitamin C is the vitamin most likely to be lost from foods as it is water-soluble and destroyed by exposure to air. In preparation, using water to wash fruits and vegetables, blanching for freezing, or standing prepared vegetables in water would all lead to vitamin C loss. Vitamin C is also destroyed when an enzyme, present in fruits and vegetables, is released when the food is cut, for example when preparing salads. Loss is accelerated when heat is applied and by the presence of certain metals, for example copper and iron. As with vitamin B-complex, foods which have been heat-processed in cans or jars will have suffered loss of vitamin C. Dehydration of foods, for example, with dried fruits, cause vitamin C loss, and its sensitivity to air also means that vitamin C is lost when fruits and vegetables are stored.

Vitamin E is also sensitive to air and can be lost with storage or long exposure.

Most other vitamins are relatively stable and loss will not be too great under normal preparation and processing conditions.

Minerals are not affected by preparation or processing.

Food manufacturers, aware of the loss of nutritional values during the preparation and processing of foods, may add nutrients back into the food as part of the processing, as for example when they add vitamins to bread products and breakfast cereals.

Question 11

How can the loss of nutrients be kept to a minimum when preparing and cooking food?

Dietary Reference Values

As well as getting the correct balance between your energy needs and diet, it is also important to ensure that, on average, the correct quantities of the other nutrients are consumed.

In the UK, the current recommended intakes for the different nutrients comes from the 1994 report of the Committee on the Medical Aspects of Food (COMA), *Dietary Reference Values for Food, Energy and Nutrients for the UK*. The report was the first to set intake levels for the major nutrients (proteins, fats and carbohydrates) as well as for vitamins and minerals. The committee also decided to replace the previous recommendations which were known as the 'recommended daily intakes' (RDI) and the 'recommended daily amounts' (RDA) with a range of values for each nutrient, known as **dietary reference values (DRVs)**.

The COMA Panel decided to set a range of DRVs in order to emphasise that the recommendations were population estimates and not recommendations for daily intakes by groups or individuals. Individual requirements for nutrients vary considerably depending on factors such as age, sex, size, metabolic rate and occupation, as well as on the rest of the diet, which may alter the efficiency of absorption or utilization of certain nutrients. The body also has stores of certain nutrients, for example fat-soluble vitamins, so variations in daily intake of such nutrients can be accommodated. Thus it could be misleading to recommend a particular set daily intake level.

The COMA Panel set four levels of DRVs:

- The level at which 50% of a population would need more of that nutrient and 50% would need less, this level is referred to as the **estimated average requirement (EAR)**
- The level that would be enough only for the few people (2.5%) in a group with low needs for that nutrient; this level is known as the **lower reference nutrient intake (LRNI)**
- The level that would be enough, or more than enough, for 97.5% of people within a certain group; at this level deficiency would be unlikely. This level is the reference nutrient intake **(RNI)** (Confusingly, RNI values are still often quoted as RDA or RDI on food labels)
- Where there was insufficient information about the human requirements for a particular nutrient, for example vitamin E, a **safe intake** level was set, which was sufficient for most people's needs but would not be too high an amount to cause undesirable effects.

Question 12

Why do think that COMA chose to recommend a range of intakes for different nutrients?

Activity 9

Look at the DRVs for your own age and gender (you'll find these in the Manual of Nutrition). Make a list of these. Now look at your intake of food for one day and try to assess whether you are within the values given. If not, how could you amend your diet so that you could achieve the targets?

It is interesting to note that the recommendations of other committees, such as those set up by the World Health Organisation and others in the USA, have arrived at values that are somewhat different to the DRVs set by the COMA Panel.

Nutritional requirements of different groups

Although the Dietary Reference Values are based on whole population estimated requirements, and individual needs vary according to a range of factors, it is still possible to generalise about the nutritional requirements of different groups within a population.

Babies

In the first months of life babies rely entirely on milk to provide all their nutritional requirements. Breast milk provides these nutrients in the correct amounts for human babies, as well as containing several natural agents which help protect the baby against disease. Cows' milk, which is processed to make it resemble human milk, provides the nutrients that babies need but not the protective agents.

At the age of about 3-4 months, the baby's store of iron is depleted, so iron-rich foods should be added to the diet, these include minced liver, pureed apricots and fortified baby foods.

Young children

As young children are growing and developing rapidly, their nutritional requirements are relatively high, especially for energy and protein. As protein and carbohydrate foods can be bulky when providing a high level of energy, young children can be given high-fat foods in order to meet their energy requirements without eating too much food, for example full-fat milk rather than skimmed, butter or vegetable margarine rather than low-fat spread. In addition, a range of vitamins and minerals are important, both to help physical development and to protect against infection and illness.

Pregnant and lactating women

When a woman is pregant or breast-feeding her nutritional requirements increase. This is not only to provide nutrients for the growth and development of the fetus but also the provide sufficient nutrients for the changes taking place in the woman's body. The fetus will require energy, protein, iron, calcium, folic acid and vitamins C and D, if these are not provided in the diet then the woman's own body stores will be used, and will become depleted. The woman will also require additional energy and iron.

Older people

There is much variation in the nutritional needs of the elderly due to their individual body changes, lifestyle changes and state of health. Some elderly people continue to stay healthy and lead active lives, in which case their nutritional needs continue to be the same as for other adults. However, if they become less active then their energy requirements are reduced, it has been suggested that energy requirements reduce by 3% from the age of 25-35 and from 35-45; by 7.5% from age 45-55 and 55-65 and by 10% from 65-75. If this is the case, then energy requirements reduce by two-thirds from the age of 25 to the age of 75. However, whilst energy requirements decline those for protein, vitamins and minerals do not. Calcium and vitamin D are particular important as bone density decreases with age. For those with physical illness the diet should provide protein, some fibre and a good range of vitamins and minerals. For those who find themselves housebound, the diet should include vitamin D to make up for the lack of sunshine. Often, poor appetite in the elderly means that they do not eat adequate amounts to provide the range of nutrients required, if this happens they should be encouraged to eat light but nutritious foods, such as eggs, milk, bread, fruits and vegetables.

Vegetarians

There are different forms of vegetarianism – a vegetarian will not eat any food from animals, whether the animal is killed or not. Many vegetarians, however, will eat milk, cheese and eggs, as the animal is not killed to provide these foods. These are called lacto-vegetarians. Strict vegetarians, who eat no foods from animals, or with animal ingredients, are vegans.

Lacto-vegetarians, providing they eat a varied diet, will be able to satisfy their nutritional needs. Vegans, however, need to plan their diet more carefully to ensure that they eat sources of vegetable protein (beans, peas, nuts, lentils, cereals) which, in combination, will provide their protein requirements (see the section 'The structure of nutrients' for an explanation of this). They also need to ensure that wholemeal products are eaten, to supply B-vitamins and iron-rich foods, such as Marmite, dried fruit and fortified breakfast cereals. Marmite is also a good source of vitamin B_{12}, which is mainly found in meat.

Creating foods with useful nutritional properties

From what you have read in this section, you should now be able to see how manufacturers develop food products to meet specified nutritional criteria. Some examples of where this has been done include high-energy/high-protein bars for sportspeople; low-energy/low-fat foods for slimmers; low-sugar foods for diabetics. Many foods now have artificial vitamins and minerals added to improve their nutritional value, for example margarine, breakfast cereals, bread. Food products are also now being produced with the claim that they can improve, or have positive effects on, health, for example Aviva and Benecol products.

Diet and health

Introduction

As the food we eat provides our bodies with the nutrients needed for growth, development and maintenance, it is fairly obvious that there is a link between the food we eat and the state of our health. Other factors are also involved, such as the amount and type of exercise taken but here we shall look at the links between food intake and health.

Diet and health

One of the major changes in diet during this century has been the increasingly wide variety of foods available to us. However, this wide choice often makes things more difficult. How do we know what we should eat, or avoid, to ensure a healthy diet? Unfortunately, it is not easy to answer this question at a general level because each of us will have individual needs. What some organisations have done, (such as the Department of Health in the UK, the National Academy of Sciences in the USA and the World Health Organisation), to try and help answer the question is to produce recommended nutrient intakes for the 'average' person. (These are dealt with in detail in 'Food and Nutrition')

Although there are some differences in recommendations, there are also a number of clear messages relating to food intake and health. The major concerns of Western countries, where the average diet is nutritionally adequate, are the diseases such as chronic heart disease and cancer, both of which are believed to be related to diet. This led the National Advisory Committee on Nutrition Education (NACNE) in the UK to make recommendations as to what might constitute a healthy diet. The main recommendations are shown in table 1.

It must be remembered that the consensus of what constitutes a healthy diet does shift from time to time, depending on what is fashionable, important, reasonable or provable! The NACNE recommendations, however, are still current so we shall look at them in more detail.

Table 1 nutritional guidelines proposed by NACNE (1983)

Dietary component	Current estimated intake	Proposals: long-term	short-term
energy intake		see note (a)	
fat intake	38% of total energy	30% of total energy	34% of total energy
saturated fatty acid intake	18% of total energy	10% of total energy	15% of total energy
polyunsaturated fatty acid intake	–	no specific recommendation, see note (b)	
cholesterol intake	–	no recommendation	
sucrose intake	38kg per head per year	20kg per head per year	34kg per head per year
fibre intake	10g per head per day	30g per head per day	25g per head per day
salt intake	8.1-12g per head per day	recommended reduction by 3g per head per day	recommended reduction by 1g per head per day
alcohol intake	4-9% of total energy	4% of total energy	5% of total energy
protein intake	11% of total energy	no recommendation	

Notes

a) recommended adjustement of the types of food eaten and an increase in exercise so that adult body weight is maintained within the advised limits of weight for height

b) in practice, there is likely to be a greater consumption of both polyunsaturated and monounsaturated fatty acids and a tendency for the ratio of polyunsaturated to saturated fatty acids to increase

Fat intake

One of the recommendations is to reduce the amount of fat in our diet and move to higher proportions of unsaturated fats, because of a possible link with coronary heart disease, cancer of the breast, colon and prostate, as well as hypertension. The term 'possible link' is used not to engender total scepticism about the scientific basis for the link but to emphasise that in such matters the connection is difficult to prove, (much more difficult than, for example, the link between scurvy and vitamin C). Figure 1 shows the relationship between deaths from heart disease and breast cancer and total fat intake as a percentage of dietary energy in a range of countries.

It is important to stress that such graphs do not establish a direct causal link. It may not be the fats themselves that are the problem, but some other factor

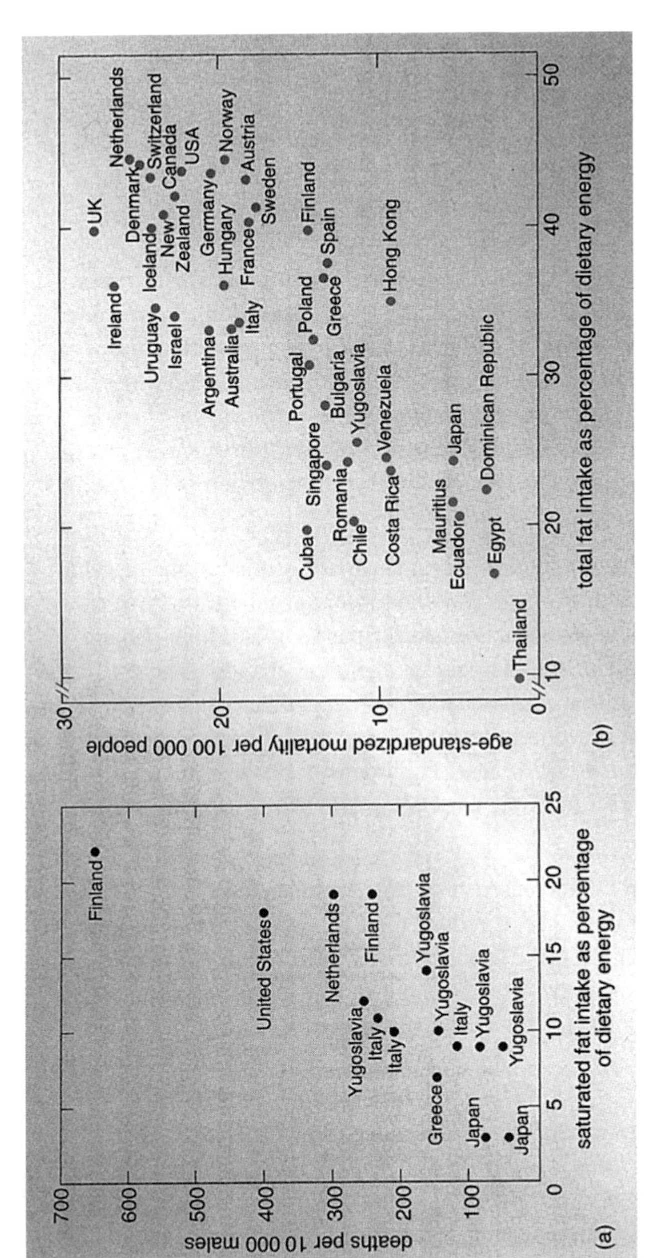
(a)
deaths per 10 000 males
saturated fat intake as percentage of dietary energy
Finland
United States
Netherlands
Finland
Yugoslavia
Italy
Italy
Yugoslavia
Yugoslavia
Greece
Italy
Yugoslavia
Japan
Japan
Yugoslavia
(b)
age-standardized mortality per 100 000 people
total fat intake as percentage of dietary energy
UK
Ireland
Netherlands
Denmark
Switzerland
Iceland
Canada
New Zealand
USA
Uruguay
Israel
Germany
Norway
Hungary
Austria
France
Sweden
Argentina
Australia
Italy
Finland
Poland
Spain
Portugal
Greece
Bulgaria
Yugoslavia
Cuba
Singapore
Romania
Chile
Hong Kong
Costa Rica
Venezuela
Mauritius
Ecuador
Japan
Dominican Republic
Egypt
Thailand

Figure 1

reflected by dietary fat intake, such as average annual income. Nevertheless, in the case of fats, the association is backed up by other studies which show that a falling consumption of saturated fats is matched by a reduction in the instances of heart disease, which is why it is recommended to reduce fat intake.

Question 1

How could saturated fats be reduced in the diet? And how could polyunsaturated fat intake be increased?

Cholesterol

Cholesterol used to be regarded as the most dangerous of all the common dietary constituents. Indeed, high concentrations of cholesterol in the blood are associated with a high risk of chronic heart disease, providing a way of predicting the likelihood of chronic heart disease. However, cholesterol in the diet is now not thought to be an important cause of this disease. There seems to be little relationship between the levels of cholesterol in our food and that in our blood, which explains the lack of any recommendation in the NACNE report.

What is known, though, is that cholesterol is carried around in the blood by proteins, this combination of cholestrol and protein is called 'lipoprotein'. There are two types of cholesterol, low-density or LDL and high-density HDL. Low-density cholesterol undergoes a chemical change when it is taken up by cells and it is this that causes plaque, which is then deposited on artery walls and builds up, so preventing blood flowing freely through the arteries. It is this which then causes heart disease. High-density cholesterol, on the other hand, removes cholesterol from the circulation and offers some protection against heart disease.

Low-density cholesterol is increased if the diet includes high levels of saturated fats. Polyunsaturated fats reduce the LDL but, unfortunately, also reduces high-density cholesterol. Monosaturated fats, found in olive oil and avocado, can help to lower low-density cholesterol without reducing high-density cholesterol, so provide the best protection of all.

Sugar

There are two problems caused by a high intake of dietary sugar: first, sucrose is used by our bodies primarily as a source of energy, thus its inclusion in our diet as a sweetener serves only to increase our energy intake, increasing the risk of obesity. Second, it can cause tooth decay, the problem increasing the longer the sugar is in the mouth. Hence, on both counts, there are recommendations to reduce our intake.

Question 2

What are the health problems associated with obesity?

Other than reducing the amount of sugar we add to food (for example in tea, coffee, on breakfast cereals) how can sugar intake be reduced?

Dietary fibre

Associations have been claimed between high fibre intake and reduced incidences of heart disease, hypertension, bowel cancer, appendicitis and piles. However, the Department of Health Standing Committee on Medical Aspects of Food Policy (COMA) said in 1991 about fibre, (which it referred to as non-starch polysaccharides)

> *That it is not currently possible to identify NSP as a major factor in the aetiology of these diseases.*

The problem is one of multiple associations. High fibre diets are by definition high in vegetable and fruits, hence they will be associated with high levels of digestible starches and be low in animal products. Thus high fibre intake is more a marker of a particular kind of diet.

However, if the diet is too high in fibre this may lead to a reduction in the absorption of calcium, iron and phosphorus. This is because the dietary fibre may 'bind' these minerals. and make them unavailable

Question 3

Dietary fibre (NSP) is not a nutrient, so in what way is it useful in the diet?

How can intakes of dietary fibre be increased?

Salt

Sodium and chloride ions are essential and are widely distributed in body fluids. We need about 4g of salt per day to replace that lost in sweating. On average, an adult consumes about three times that amount, the surplus being excreted in the urine. There is good evidence that there is an association between high salt intake and the development of high blood pressure. It has been shown that reducing salt intake does help to lower blood pressure in those people who already suffer from hypertension. However, other studies have shown that this association is only apparent for about 10-20% of the population who show a special susceptibility to salt.

Question 4

Apart from salt itself, which foods are considered to be 'high-salt' foods?

Alcohol

Drinks in the diet are mainly water. The nutritional value of tea, coffee, carbonated drinks and squashes is minimal, but they do provide an important means of taking in water. The exception to this is alcohol, which also provides energy. The average bottle of wine contains about 2,200kJ of energy; even if you drink only half of it, this is still equivalent to about 10% of your average daily energy requirement! There are also other effects related to absorption of alcohol by the body, which can have detrimental effects on the liver. NACNE, therefore, recommended a reduction in alcohol consumption. There is now, however, other evidence which indicates that alcohol in moderate amounts exerts a protective effect against chronic heart disease.

Protein poisoning

Some recent research seems to be indicating that it is possible to have too much homocysteine in the blood. Homocysteine is a protein amino acid, which is found in meat, milk and eggs, and which the body uses to build tissue. The research found that homocysteine can also injure blood vessel linings, hasten the build-up of scar tissue and encourage blood clots. As well as diet, smoking and low activity levels can increase homocysteine in the blood.

Ways to reduce homocysteine levels would be to reduce intake of meat, milk and eggs, and also increase B-vitamin intake, especially B_6, B_{12} and folic acid. These B-vitamins help to break down excess homocysteine in the blood.

Research into the scientific and nutritional aspects of foods continues to provide us with more information, some confirming what was known and some providing new insights. It is important to realise that this information never stands still, it continues to develop, and you should attempt to keep abreast of new developments.

References

Committee on Medical Aspects of Food Policy 1991, *Nutritional Aspects of Cardiovascular Disease: Report of the Cardiovascular Review Group*, HMSO.

Committee on Medical Aspects of Food Policy 1994, *Dietary Reference Values for Food, Energy and Nutrients for the UK*, HMSO.

Department for Education and Employment/Qualifications and Curriculum Authority 1999, *The National Curriculum Handbook for secondary teachers in England,* DfEE/QCA.

Design and Technology Association 1995, *Minimum Competences for Students to teach Design and Technology in secondary schools*, DATA.

Logue, A.W. 1991, *The Psychology of Eating and Drinking*, W.H. Freeman.

National Advisory Committee on Nutrition Education 1983, *Proposals for nutritional guidelines for health education in Britain*, Health Education Council.

Times Educational Supplement 28.04.00 Briefing p.21

Walker, R., Dobson, B., Middleton, S., Beardsworth, A. and Keil, T. 1995 'Managing to eat on a low income' in *Nutrition and Food Science* 3 pp. 5-10.

Further Reading

Fieldhouse, P. 1998, *Food and Nutrition*, Stanley Thornes

Fox, B.A. and Cameron, A.G. 1995, *Food Science, Nutrition and Health*, Arnold

Gaman, P.M. and Sherrington, K. 1996, *The Science of Food*, Butterworth-Heinemann

Piper, B. 1999, *Diet and Nutrition*, Stanley Thornes

Shaw, R. (1996) *Product Development Guide for the Food Industry*, Campden and Chorleywood Food Research Association Chipping Camden (available from Ridgwell Press, P.O. Box 3425, London. SW19 4AX)

Answers to Questions

Influences on food choice

Question 1

Those making food choices based on nutritional influences are likely to include those with some knowledge of nutrition, such as dieticians, food scientists, home economists. Others may include those with food-related illnesses, such as diabetics, coeliacs, those with high blood pressure or heart disease. Sports people may choose food for its particular nutritional qualities, vegetarians and those with a general interest in maintaining good health may also have a knowledge of nutrition which they apply.

Question 2

Lack of money to buy food usually results in low levels of food consumption. In addition, the types of food bought are usually those with a high-fat, high-sugar content because they are most satisfying and have little waste. There is usually a low consumption of wholemeal products, fruits and vegetables as these are perceived to be more expensive, or do not provide the energy required.

Question 3

At different times of our lives we respond differently to social influences. Young children will be influenced by their family, for example if mummy or daddy is seen to be enjoying a food the child may well be influenced to copy them. Other influences may come from television advertising, particularly where the food is linked to a children's television character or famous personality, for example Barbie pasta shapes.

Adolescents are likely to be highly influenced by their peers, this means that certain foods or eating patterns may be part of their social life, for example meeting friends at McDonalds. They may also be influenced by 'trends', for example many adolescents, particularly girls, experiment with vegetarianism. A current concern, again particularly with girls, is the trend for models to be extremely thin thus causing adolescent girls to reduce their food intake to low levels in order to achieve this look.

Women, too, may also be influenced by the look of models and constrain their food intake in order to achieve what they consider a desirable figure. Some women may also be influenced by trends, for example for low-fat eating; high fibre diets; the Hay diet or other acclaimed diet. Women with families may be influenced by their likes, dislikes and needs.

Some men, too, are concerned with their appearance and health and may be influenced by social trends. Others will be influenced by peers and others by family needs.

Question 4

Some years ago technology changed the British diet with the introduction of frozen foods and canned foods. Developments in food processing made a wider range of foods available, for example Quorn (a meat substitute). Changes in transport and storage technology made many more foods available, foods from other countries can now be easily obtained, as can foods previously only available for short seasons during the year. Irradiated foods can be kept fresh for longer. Genetically modified foods promise bigger and better crops.

Developments in packaging technology had an impact, for example fresh soup in cartons.

Changes in domestic technology also had an impact, the introduction of the microwave led to increasing consumption of ready-made meals.

Question 5

Food manufacturers have been adding nutrients to foods for some time now, but recent developments have seen the promotion of 'health-giving' foods. Foods are being developed which claim to be beneficial to the heart, the bones and the digestive system.

Food product development

Question 1

Influences include the development and availability of new materials and processes; competition between food companies; fashion trends, e.g. those set by top restaurants; TV and media; celebrity chefs; recipe books; health and government guidelines.

Question 2

Appearance: descriptors relating to colour, shape, size, e.g. colourful, glossy

Texture: crunchy, smooth, crisp, chewy, short, juicy, crumbly

Aroma: savoury, pungent, stale, yeasty, fruity, spicy

Taste: sweet, sour, salty, acid, bitter

Question 3

These are critical:

To ensure safety and quality

To try out ideas in practice (to check their feasibility)

To ensure that product ideas can be replicated in volume

To ensure that products meet the needs and preferences of consumers

Question 4

Imaging and modelling can be done:

With small quantities of ingredients and small production runs

using spreadsheets to model nutritional analysis, costing or to work out ratios, proportions and formulations (mathematical modeling)

By drawing and sketching

Using a computer software program to model and predict bacterial growth when calculating shelf life.

Food Manufacturing

Question 1

For the manufacturer:

Eliminates the need to store expensive stock

Keeps costs down

Transportation of stock becomes the responsibility of the supplier

Saves on space

Reduces waste

For the customer:

More responsive to customer needs and demands

Keeps costs, and so prices, down.

Question 2

- Raw materials are checked for quality on delivery
- Raw food handlers prepare chicken and remove bones
- Marinade recipe is mixed (yoghurt, herbs, spices, water)
- Chicken is marinated to absorb flavours
- Chicken pieces racked for cooking
- Chicken is cooked (must reach >80°C)
- Product passes through to separate cooked food handlers
- Chicken is rapidly chilled (<4°C)
- Dish is packaged and wrapped
- Dish passes through metal detector
- Product placed in cold storage (0 to +2°C)
- Distributed in refrigerated lorries to retail outlets.

Question 3

Heating would be carried out:

If a product needs to be cooked, e.g. to improve its appearance, texture, taste or aroma

To change the properties of ingredients in order to process them, e.g. to make something liquid in order to deposit it

To improve keeping qualities and extend shelf life, by destroying micro-organisms.

Cooling would be carried out:

To reduce temperature after heating or cooking

To reduce the rate at which micro-organisms multiply

To halt microbial development

To set a mixture

Question 4

Benefits:

Improved nutritional value, e.g. foods with higher protein content

Improved keeping qualities and shelf life

More disease-resistant crops

Improve desirable characteristics and reduce undesirable ones

To diagnose whether crops need treatment, reducing unnecessary treatment

Drawbacks:

Concerns over long term health and safety effects

possible detrimental effects on the environment.

The chemical structure of nutrients

The chemical structure of nutrients

Question 1

Essential amino acids need to be included in the diet as the body cannot product them for itself. Non-essential amino acids can be made by the body so do not have to be ingested.

Question 2

Some protein foods do not contain the essential amino acids, or have insufficient quantities for human needs. If a protein food low in one essential amino acid is eaten at the same time as another protein food, which is low in a different amino acid, but a good source of the amino acid missing in the first food, they can combine in the body to provide the full requirement of the essential amino acid.

This would be useful in vegetarian diets, where peas, beans and cereal foods can be combined. Also, in diets where there was insufficient money to provide sufficient animal protein foods, cheaper vegetable sources of protein can be eaten.

Question 3

The sequence in which amino acids join together is one determinant of the protein shape. The way in which the amino acids then form cross-bonds will further determine its shape, for example those with more non-polar R groups will tend to form tight clusters whilst those with fewer will be more open.

Question 4

R group

valine: $(CH_3)_2CH{-}C(H)(NH_2){-}COOH$ — R group: $(CH_3)_2CH{-}$

leucine: $(CH_3)_2CH{-}CH_2{-}C(H)(NH_2){-}COOH$ — R group: $(CH_3)_2CH{-}CH_2{-}$

isoleucine: $CH_3CH_2(CH_3)CH{-}C(H)(NH_2){-}COOH$ — R group: $CH_3{-}CH_2{-}CH(CH_3){-}$

Question 5

Hydrogen atoms affect protein shape by forming bonds with oxygen atoms which causes a folding in the protein chain. Sulphur atoms form bonds between each other, forming a bridge which holds the protein shape.

Question 6

A change of jut one amino acid in the sequence of a protein chain would affect its structure because it may cause a hydrogen-oxygen bond to be formed or broken; it may cause a disulphide bridge to be formed or broken or it may affect the clustering of R groups. This is because the changed amino acid may have its hydrogen and oxygen atoms in different locations, it may or may not contain sulphur and, again, this may be differently located and it may contain a polar or non-polar R group.

Question 7

Fibrous protein, when heated in water, allows water molecules in. This causes the hydrogen bonding in the protein chain to weaken and softens the structure of the protein, making the food easier to eat. Globular proteins, when heated, denature. This means that their structure 'unravels' and reforms in a new structure. This change is permanent, the original structure cannot be reformed.

Question 8

The sugar unit found in all sugars is glucose. One glucose unit joined to another glucose unit forms maltose. One glucose unit joined to a galactose unit forms lactose.

Question 9

Cellulose cannot be digested by the human body because the way in which the long chains of glucose units join together cannot be broken by the human digestive system. This means that the cellulose cannot be broken down into glucose units for absorption into the body.

Question 10

Lard is mostly saturated fat, as it is an animal fat and is very solid.

Question 11

Cornflour contains starch which gelatinises when heated with water, this will cause the mixture to thicken. Butter adds colour and flavour. The lemon juice, lemon rind and sugar add flavour. The protein in the egg yolks will denature when heated and will cause the mixture to hold its shape as the lemon filling cools.

Digestion of Food

Question 1

Peptidases will break down the peptide chains of protein. Amylases will break down the long chains of amylose which make up starch (carbohydrate). Maltases will break down the maltose chains of sugar into its constituent glucose units. Lipases will break down fats into glycerol and fatty acids.

Food and Nutrition

Question 1

there is extra need for protein during childhood and adolescence, when the body is growing and developing rapidly. Pregnant women need additional protein to provide for the growing fetus. During some illnesses, protein will help recovery as it will aid the repair of tissue and growth of new tissue.

Question 2

Non-starch polysaccharide is important in the diet because it aids the efficient excretion of waste from the digestive system.

Question 3

Digestible starch can be found in cereals, cassava, yam and potato. Intrinsic sugars can be found in sweet foods such as honey and fruit. Extrinsic sugars are those added to foods to sweeten them, for example in chocolate and jam.

Question 4

Too much fat in the diet can be harmful because it provides high levels of energy which can lead to obesity and associated health problems.

Question 5

Vitamins most commonly added to manufactured foods include the B-group, as these are often removed from foods during processing, for example when the wheat bran is removed to make white flour. Fat-soluble vitamins A and D may also be added to margarine, as they would occur naturally in butter.

Question 6

Fat-soluble vitamins, particularly vitamin A, can be dangerous if too much is consumed.

Question 7

Flour, bread and breakfast cereals often have calcium added, this is because calcium is removed during processing. Salt is also added to most processed foods as it brings out the flavour of the food.

Question 8

Dietary fibre is essential for good health because it causes the efficient removal of waste products from the body, so keeping the digestive system healthy.

Question 9

Roast beef provides a good source of protein and fat, vitamin B_{12}, nicotinic acid, some calcium iron and zinc. Yorkshire puddings, made from eggs, milk and flour, would provide protein (from eggs and milk), fat (from eggs and milk), carbohydrate (from flour), calcium (from milk and flour, if fortified), vitamin A (from milk), sulphur and phosphorus (from eggs) and calcium, thiamin and nicotinic acid if the flour was fortified. If wholemeal flour was used there would also be some B-vitamins, vitamin E, iron and dietary fibre. Potatoes would provide carbohydrate, if fresh they may provide some vitamin C, if roasted they would provide fat and if cooked with skins on they would provide dietary fibre. Carrots are a good source of vitamin A, peas would provide some vegetable protein and some vitamin C if fresh. Gravy would provide carbohydrate and, if made using water from the vegetables, some B-vitamins and vitamin C.

Question 10

An individual's need for energy is affected by a range of factors including gender, size, age, metabolic rate, level of activity and state of health.

Question 11

To keep nutrient loss to a minimum when preparing and cooking food it is important to prepare the food just before cooking (do not leave it standing); use minimum amounts of water; retain the skin where possible; cook for the least amount of time; avoid using copper or iron pans; serve as soon as possible after cooking; use cooking water where possible, for example for gravy or soup.

Question 12

COMA recommended a range of intakes for different nutrients because individual requirements depend on a range of factors. The recommendations, by covering a range of intakes, covered the needs of different groups within a population.

Diet and health

Question 1

Saturated fats can be reduced by reducing the overall intake of animal fats, such as cheese and meat. They can be further reduced, and polyunsaturated fats increased by replacing other animal fats with those from vegetable sources, for example replacing butter with vegetable margarine spreads. When frying foods, vegetable oils can be used in place of lard or butter. Using vegetable oils in salad dressings will also increase polyunsaturated fat intake.

Question 2

Obesity can lead to heart disease and hypertension. Sugar intake can be reduced by buying processed food (such as breakfast cereals and tinned fruit and vegetables) without sugar added; by eating fresh or dried fruit in place of cakes, biscuits and chocolate and by replacing sugar with sweeteners, such as aspartame. (If changing to a chemical sweetener, check that there are no allergic reactions).

Question 3

Dietary fibre is useful in the diet because it encourages the efficient elimination of waste from the digestive system. Dietary fibre intake can be increased by eating fresh fruits and vegetables, with skins on where possible, and by eating 'wholegrain' or 'wholewheat' breakfast cereals or porridge oats. Changing white flour and white flour products, such as bread, cakes, biscuits and pasta, to wholemeal flour products will increase dietary fibre, as will eating brown rice in place of white rice.

High intakes of dietary fibre, however, can have a detrimental effect as it may cause a reduction in the absorption of calcium, iron and phosphorus.

Question 4

High-salt foods are generally manufactured foods, such as tinned vegetables; sauces and ketchups; breakfast cereals; ready-made and savoury snack foods. Other foods high in salt are bacon, ham, sausages and other processed meats; butter; cheese and bread.